SOMEONE IS HIDING SOMETHING

What Happened to Malaysia Airlines

FLIGHT 370?

SOMEONE IS HIDING SOMETHING

What Happened to Malaysia Airlines

FLIGHT 370?

RICHARD BELZER,

GEORGE NOORY,

AND

DAVID WAYNE

Skyhorse Publishing

SOMEONE IS HIDING SOMETHING

Skyhorse Publishing books may be purchased in bulk at special discounts for sales promotion, corporate gifts, fund-raising, or educational purposes. Special editions can also be created to specifications. For details, contact the Special Sales Department, Skyhorse Publishing, 307 West 36th Street, 11th Floor, New York, NY 10018 or info@skyhorsepublishing.com.

Skyhorse® and Skyhorse Publishing® are registered trademarks of Skyhorse Publishing, Inc.®, a Delaware corporation.

Visit our website at www.skyhorsepublishing.com.

10 9 8 7 6 5 4 3 2 1

Library of Congress Cataloging-in-Publication Data is available on file.

Cover design by Brian Peterson

Cover photo © Rob Griffith/AP Corbis

Print ISBN: 978-1-63220-728-9
Ebook ISBN: 978-1-63220-850-7

Printed in the United States of America

To the families and friends of those on Flight 370:

The truth is sometimes elusive. The important thing is that somebody is searching for it. This book attempts to document the facts and possibilities known at this time. May they assist you in that endeavor.

To the 239 victims and their families:

You are not forgotten.

"In short, a deductive argument must be evaluated in two ways. First, one must ask if the premises provide support for the conclusion by examining the form of the argument. If they do, then the argument is valid. Then, one must ask whether the premises are true or false in actuality. Only if an argument passes both these tests is it *sound*."[1]

—Internet Encyclopedia of Philosophy

"If you can't solve a problem, enlarge it."[2]

—Dwight D. Eisenhower

Table of Contents

Introduction

Transparency has been consistently absent from the authorities' public assessment of what might have happened to Malaysia Airlines Flight 370. The authorities seem to be hiding the full story from the public and, most importantly, the families of the victims. If they had any shred of human decency, they would reveal every known fact about Flight 370 and every plausible explanation for why it never made it to its intended destination. Mainstream media has also done an abominable job. They've misled the public by making reports without investigating the facts.

The families need and *deserve* answers—and so does the public. If a jetliner can simply "disappear"—and in the presence of considerably sophisticated technologies—what does that say about the world we live in? What's to stop it from happening again?

There are many things the media hasn't told you about. For example, did you know that the families of those on board Flight 370 formally wrote and demanded the Malaysian authorities answer questions regarding the plane's disappearance? Did you know that those questions were more intelligent and relevant to the plane's disappearance than the ridiculous things asked by mainstream media? But have you heard anything about those questions? Probably not.

It's been many months since Flight 370 went missing, yet our questions and the victims' families' questions continue to go unanswered. Among a great many other things, *we need to know what happened to the plane and where it is*. Officials theorize that it crashed into the ocean, but if so, *where are the parts?* The plane, a Boeing 777-ER, contains *three million* parts.[3] You would think that if it crashed, they'd have found *something*. Where are the pieces of the plane? Where are the seat cushions? The flotation devices? The luggage? Either, mysteriously, nothing has been found or pieces have been found and no one's admitted it.

Also, *why did the plane veer sharply off course?* Why did it *keep* flying off course? *How did it go off radar* and how did it stay off radar if, indeed, everyone was desperately looking for it, as the authorities claim? And *why didn't the emergency beacons, the plane's tracking devices, function?* They are designed to emit emergency signals on impact, even at the bottom of an ocean. So why didn't they?

One big thing missing in this case is the CRIME SCENE. There *isn't* one, at least not a publicly known one. One can almost picture the perpetrators thinking "*What* crime? [Insert laughter.] There won't even *be* a crime scene." But *there's no such thing as the perfect crime* because there's always a mistake, something that doesn't add up. That's what this book is about, all the things that don't add up.

The title of this book, *Someone Is Hiding Something*, is a statement that conveys the overriding international sentiment surrounding the mystery of the missing passengers and jetliner. The line was put forward by none other than a high-ranking Malaysian official, a person in a perfect position to offer an intelligent observation, the former Prime Minister of Malaysia Dr. Mahathir bin Mohamad. Fed up with the media dance and obvious obfuscations in the case, Dr. Mohamad posted a blog entry in which he blatantly accused the CIA of knowing more about the plane's disappearance than they've been willing to share. We're not making any of this up, folks. It sounds like a far-fetched movie—but no, it's all real! In eleven, straightforward points under the header "What Comes Up Must Come Down," Dr. Mohamad pointed out the obvious, but understated point that: **"airplanes don't just disappear**. Certainly not these days with all the powerful communications systems, radio and satellite tracking and filmless cameras which operate almost indefinitely and possess huge storage capacities."

Dr. Mohamad's words were not as widely publicized as they should have been, but those who read them recognized that, unlike what is being spewed out on most of network television, they make a great deal of sense. Not only can a plane not disappear, but the plane in question *especially* shouldn't be able to. He wrote, "MH370 is a Boeing 777 aircraft. It was built and equipped by Boeing. All the communications and GPS equipment must have been installed by Boeing. If they failed or have been disabled, Boeing must know how it can be done. Surely Boeing would ensure that they cannot be easily disabled as they are vital to the safety and operation of the plane."[4] "… For some reason, the media will not print anything that involves Boeing or the CIA … It is not fair that MAS and Malaysia should take blame."[5]

Dr. Mohamad wrote that even if terrorists had attempted to breach the flight deck, the plane was equipped so that it could be taken over "remotely by radio or satellite links by government agencies like the Central Intelligence Agency."[6] He wrote, "*The plane is somewhere … maybe without MAS [Malaysia Airlines] markings … Someone is hiding something.*"[7]

Well, *gee whiz*. We didn't hear about any of that on TV. Did you? This is pretty crazy and it's not like Malaysia is a third world country! Malaysia is a highly developed nation with a population of more than thirty million people, placing it as the forty-third most populated country in the world.[8] Its capital, Kuala Lumpur, with the distinctive Petronas Towers defining its skyline, is recognized as one of the most modern cities on the planet. It's one of the world's "40 Biggest Economies"[9] and boasts a sophisticated infrastructure. Its air systems, both military and commercial, are also sophisticated, equipped with very modern aircraft and high-tech radar systems.

This book will explore what could have happened to Malaysia Airlines Flight 370 through a process of *logical deduction*, to determine what fits in the equation and what does not. We will go through all the commonly discussed theories of what might have happened to the plane, as well as topics rarely mentioned in mainstream media, such as the "cyber jacking" of an airliner, which many people believe is not nearly as far-fetched as it may at first seem.[10] This book will sort that all out.

Aviation experts report that it is virtually impossible that all tracking systems on a modern passenger jet would fail unless there was a catastrophic event.[11] There are simply too many methods of tracking today. Even back-up systems have back-up systems. For this reason, there's an old adage used among air traffic controllers that "Planes want to be seen."[12] They don't disappear, because they *can't* disappear. The sophisticated technologies tracking them today are simply too good.

So, while major media reported that with the plane's transponder turned off, "the plane is flying blind from the ground's point of view,"[13] that quite simply isn't possible. The plane would be far from invisible. Even if the transponder was turned off, "there are several separate tracking and surveillance technologies and networks that feed information to air traffic controllers and provide status updates for each airplane within a controller's sector. These reporting systems involve the use of radar, satellite data links, transponders, VHF short-range radio, and long-range high frequency radio systems that allow for voice communications of position reporting."[14]

So the notion perpetrated in the media that a plane "disappears" from tracking when the transponder is turned off is patently false.

In addition to the aforementioned tracking systems, flights are also tracked by military radar. "Flight crews may also communicate by voice through aeronautical radio networks or via satellite phones on [the networks] Iridium or Inmarsat."[15] There are so many highly sophisticated forms of tracking and communicating with a major jetliner today that *only a catastrophic event could cause them all to fail.*[16]

We were also surprised to learn that highly relevant eyewitness reports were largely ignored by mainstream media. Why would they do that? It's true that police investigators are rightfully cautious about eyewitness testimony, but that's due to the fact that it tends to vary widely. When there's a common thread, investigators are trained to note that commonality and pursue that thread. If two eyewitnesses at a crime scene both say that they saw a short white guy with a knife, then it doesn't take Sherlock Holmes to figure out that maybe you should be on the lookout for a short white guy with a knife. Oddly, mainstream media and the "investigators" of the disappearance of Malaysia Airlines Flight 370 soundly *ignored* the common thread in the eyewitness testimonies, specifically the course the plane most likely took, including where it was on the night of March 8, and when it was there.

Then there was the tremendous coincidence that *another* Malaysia Airlines flight had gigantic trouble a few months later—it was shot down out of the sky over the Ukraine—and it was *also* a Boeing 777. What was particularly disturbing when that happened was the vast differences in mainstream media's coverage of it and their coverage of Flight 370. How come they immediately knew *everything* in that case of the second plane but when it came to Flight 370, they just shrugged their shoulders like a kid on an elementary school playground? *Sorry. Still can't find it.*

Being aware that something big is going on here does not stop us from feeling sensitive to the families of those who were lost. Our hearts go out to them. It's a terrible situation. But that doesn't mean that a genuine search for the truth of the case should not be pursued. In fact, because there were victims and because those victims have families who deserve answers, *the truth should be pursued aggressively*; this is not something to question and forget about.

Now, this would not be a *real* introduction to the book you are reading if we didn't say a few choice words about the role of mass media:

Walter Lippmann wrote a book called *Public Opinion* that became a classic analysis of the role of media in democracy. It was published in 1922 and is considered a seminal work because it redefined an active role for media to take in the shaping of our democracy, by intentionally using means like propaganda to *create the consent of the public*. Sound familiar? Just think "weapons of mass destruction" to allow the war in Iraq. Oh, wait, that's right, *there weren't any weapons of mass destruction.* There were weapons of *mass deception,* and that's where *creating consent* plays its role. Lippmann wrote:

"The creation of consent is not a new art. It is a very old one which was supposed to have died with the appearance of democracy. But it has not died out. It has, in fact, improved enormously in technique, because it is now based on analysis rather than on rule of thumb. And so, as a result of psychological research, coupled with the modern means of communication, the practice of democracy has turned a corner. A revolution is taking place, infinitely more significant than any shifting of economic power."[17]

As Edward S. Herman and Noam Chomsky documented in their classic work, *Manufacturing Consent,* many years later, that same media process referred to by Lippmann expanded rapidly into a system where the dominant mass media is subsidized and controlled by large bureaucracies acting in their own self-interest rather than the public good.[18] In other words, they're all in cahoots and have been for a long time. Major media has been co-opted by business and governmental interests who control it to further their aims.[19] In Chomsky's book *Media Control,* he described how they used to use "Cold War anticommunism" in the same way that they now use the "War on Terror" "as the major social control mechanism."[20]

If you considered propaganda the sole work of communist dictatorships as they taught us back in school, please think again, my friends. *Consent is manufactured.* Say it three times, then click your heels, look *away* from your television screen, and try, despite what they want you to do, to actually think for yourself.

The good news is that huge numbers of people are doing precisely that. In fact, our research for this book led us to an astounding discovery: While the explanations offered in mainstream media for the disappearance of Flight 370 are perplexing, the reactions of a curious public have been intelligent and refreshing. We poured over hundreds of articles and commentaries in mainstream media that seemed to fail miserably at addressing major points regarding the disappearance of the plane and its passengers. However, the comments from readers were asking the right questions and pointing out major inconsistencies in the officially reported version of the event.

For example: What about all those parts on a jetliner that *float?* The safety video at the beginning of every flight points out numerous items that are specifically referred to as *flotation* devices. If the plane crashed into the ocean, shouldn't they be floating out there somewhere, easily visible? If our spy satellites can find a baseball in the middle of a desert, why can't they find *those?* When the tsunami struck Thailand, items as big as refrigerators and Harley-Davidson motorcycles washed up on the shores of beaches. Are they seriously telling us that something as big as a Boeing 777 jetliner with 239 people and all of their luggage on board really crashed into the ocean and didn't leave any debris? They can't find *anything?* The ocean floor is visible to modern technology and we're supposed to believe that nobody can spot a big Boeing 777 jetliner? Come *on* now, *really?* Those are the types of cogent observations that came, and that *keep coming*, from ordinary citizens.

The fact that people ask these questions is good news for the future of the human race. People seem to innately realize it when they're being hoodwinked by The Powers That Be. The same thing occurred after President Kennedy was assassinated. If you weren't around in 1963, you may not be aware that shortly after the president of the United States was murdered in broad daylight on a Texas street, his purported assassin was conveniently eliminated, *while in police custody,* in a scenario that stunk to high heaven of foul play. That man was probably the first accused political assassin in history to deny any involvement in the political act he'd supposedly committed and he has also been convincingly linked to the US intelligence community.[21]

The public was up in arms at the ridiculous implausibility of it all and at the loss of President Kennedy under such bizarre circumstances. And then the Warren Commission, which was entrusted with investigating the assassination, did everything that they could to look the other way. It was so blatant that one historian, Walt Brown, even entitled his book on the subject, *The Warren Omission.*[22] And the media backed them up all the way, convincing everyone who owned a television that to even *utter* the word "conspiracy" was tantamount to being an un-American, commie-loving nut job who should be locked away somewhere so as not to interfere with the cocktails and golf games of the true and honorable, God-fearing, argyle sweater-wearing, proud American citizens.[23]

So, we could point out that there is a long and continuing tradition of obfuscating the truth in matters of great public import, but as a wise man once said, "Isn't it sort of redundant? Like blindfolding Earl Warren or tying Gerry Ford's shoelaces together?"[24]

Take close note though of the *reason* that many young people are unaware of that history. It's because mainstream media has conveniently curtailed a proper focus on those facts. The truth of the matter has been obfuscated, seemingly intentionally, with endless streams of mindless drivel about how there was *no conspiracy* and the ongoing implication that the glaring inconsistencies are all just a big, mind-boggling coincidence. But the public didn't buy it that time. Much to their credit, the public doesn't seem to be buying it this time either. And the former prime minister of Malaysia sure as hell isn't buying it. Dr. Mohamad put it best, and in the plainest of terms: "Airplanes don't just disappear."[25]

One thing we noticed continually in this case is that the observations of normal people far exceeded the logic in the official version of what happened. For example, as Sarah Bajc, the girlfriend of Philip Wood, one of the passengers on Flight 370, put it, it simply is not credible that the plane avoided radar after it flew off its route:

"That airplane flew, for a very long time, over Malaysian air space," she said. "You've got a 777, an unidentified object that theoretically has no communication with the ground—flying over their air space. And they're saying that their military just didn't see it—right? Or they didn't think it was a threat—they thought it was friendly. I don't believe that. The Malaysian military is quite sophisticated—they've got one of the best radar coverage systems in this part of the world. They're clearly hiding something ... But I refuse to believe that the Malaysian military ignored a 777 in their air space ... The jet had actually been accompanied by fighter planes for some period of time and there was some witness to that ... That actually makes a lot more sense than a flight being ignored by the military. So I think we need to have better view into where that plane ultimately went and who's got it now ... I'm sure that the military in Malaysia knew that that plane was there, and has tracked it in some way. Now whether they were in control of it or not, I don't know ... But the general thinking of the families involved ... they all believe this is actually a military operation of some sort."[26]

One

The Boeing 777-200ER

Everyone who's ever been on a modern jetliner like the Boeing 777 knows that they are very fragile and, for example, if you even flush a napkin down the toilet in one of those tiny little restrooms it could cause a problem so huge that the whole damn plane could come crashing down. It even says that on a sign in the lavatory, right?

Wrong, people. Signage aside, we have been seriously misled.

Trust me, you could flush the New York Yankees down that toilet and the plane would keep traveling along just fine, thank you very much.

Let's talk about the specifics of the Boeing 777-200ER. The "ER" stands for extended range. This plane was the first aircraft in aviation history to earn FAA (Federal Aviation Administration) approval to fly extended range twin-engine operations since it went into service in 1995.[27] The 777 can cruise at altitudes up to 43,100 feet and has become the flag-ship of many airlines due to its excellent track record.[28] The 777 has flown almost five million flights[29] and is universally regarded as an excellent aircraft with the latest technologies commercially available.

The 777 also has a next-generation system that is highly significant: the "fly-by-wire" system for primary flight controls. Fly-by-wire is a relatively new automated system that was once only used by advanced military jets. It is a particularly precise method in which the multiple computer systems of the aircraft actually control the actions of the plane.[30]

For obvious reasons, we are going to focus on the following aspects of the plane:

EMERGENCY EQUIPMENT ON THE BOEING 777-200ER

Communications Systems

There are many methods of communication between the air-plane, its pilots, its manufacturer, air traffic control (ATC), satellite functions, normal radar, and military radar. The meth-ods are independent of each other so that if one fails there are always other options for communication from the plane to ground personnel and for their tracking of the aircraft.

Primary radio contact is via normal radio communications between the flight deck and air traffic control. It is high fre-quency (HF) and very high frequency (VHF). If ATC suspects that the flight crew is unable to respond via radio frequency, ATC will instruct them to "squawk their IDENT." That's accomplished by the simple push of a button in the cockpit that lights up a signal for the controllers at ATC. Squawking the IDENT of the plane is such a basic emergency communications procedure that all flight crews know how do it.[31] Satellite communications systems (SATCOM) are also standard state-of-the-art equipment on the Boeing 777. And selective calling (SELCAL) is another tool the aircraft has to communicate via a unique (or virtually unique) code for each flight.[32]

Contrary to what you may have heard from some of mainstream media, commercial flights *are* tracked over an ocean. Air crews should *always* be in contact with both air traffic control and company dispatchers on the ground. Most intercontinental aircrafts even have data link or satellite communications systems that allow for constant real-time tracking."

Veteran commercial pilot Patrick Smith, the author of *Cockpit Confidential*, provides a description on his website, Ask the Pilot, of how those functions work on a flight over the ocean:[33]

> To be clear, planes *are* tracked over the ocean, even in remote, non-radar areas. This is something the media hasn't been good at explaining.
>
> It's a semantic discussion to some degree (what does 'tracking' mean?), but headlines the likes of "U.N. to Consider Ways to Track Planes Over Seas," and similar phrasings, which have been rampant, give people the impression that once a plane hits oceanic airspace, it effectively disappears until making landfall on the other side. This is *not* the case, at all.
>
> Crews are *always* in touch with both air traffic control *and* company personnel on the ground, and both of these entities are following and tracking you. Transponders aren't used in non-radar areas, but you've also got HF radio,

SATCOM, CPDLC [controller-pilot data link communications], FMC [flight management computer] datalink and so forth. Which equipment you're using to communicate depends where you are and which air traffic control facility you're working with.[34]

Transponder

The transponder is a small box that, at regular intervals, emits an electric signal. It "squawks," sending a burst of data that reveals information about the plane and its location.[35]

A former FAA safety inspector, David Soucie, the coauthor of *Why Planes Crash*,[36] was asked about Flight 370 and the Boeing 777 model, and if making it "invisible" could be accomplished by something as simple as the pilot turning off the transponder. Here's how he responded:

> Everything on that airplane is triple redundant. The electrical systems, the charging systems, the battery systems, the communications systems, even the transponders are on completely separate busses. The chance that all the electrical system was out of that aircraft would have indicated a much more massive failure of some kind.[37]

So, although you may have heard, as reported by various news sources, that the transponder of Flight 370 was turned off,[38] that simply has not been proven. We do not have a way of knowing if the transponder was actually turned off; we only know that it ceased to function in the manner in which it was designed to work.[39] That could have resulted from an immediate physical catastrophe, like a missile hit massive fire, or because it was turned off manually. We simply don't know. As Patrick Smith describes it:

> With respect to the missing Malaysia Airlines plane, a discussion of transponders is only partly relevant in the first place. For air traffic control purposes, transponders only work in areas of ATC radar coverage. Once beyond a certain distance from the coast, the oceans are *not* monitored by radar, and transponders are *not* used for tracking. We keep the units turned on because the TCAS anti-collision system is transponder-based, but we rely on SATCOM, ACARS [aircraft communications and reporting system], FMS datalink, and other means for position reports and communications. Thus transponders are pertinent to this story *only* when the missing plane was close to land. Once over the open water, on or off, *it didn't matter anyway*.[40]

Don't misunderstand, a transponder *can* be turned off. Reports of that are usually referring to what is known as ACARS being turned off, but ACARS is one—and *only* one— of an intentionally redundant system of available methods of communication with an aircraft. Another system called the automatic dependent surveillance manager (ADS) also automatically sends position reports of the aircraft. Its reports go to both air traffic control and the manufacturer—Boeing in the case of Flight 370. Like ACARS, the ADS system can be turned off manually, but it too continues to communicate via satellite when turned off. Now, having said this, note that all the following safety and emergency locator systems cannot be turned off. They function automatically and independently of each other. No one has to enable *anything* in order to make that happen because "aircraft communications systems automatically use satellites."[41]

Further delineated below are some of the extensive security systems of a Boeing 777:

Emergency Locator Transmitters

Contrary to popular misconception, there is *not one* emergency locator transmitter (ELT) on a 777, *there are three*. One is mounted on top of the aircraft above the passenger cabin and the other two are located on the large exit doors. The ELTs are "armed," meaning that they are automatically activated by extreme circumstances, such as high-deceleration forces like a crash or even a safe, but hard, landing. The ELTs can also be activated manually by the flight crew in the event that help is needed. The crew would simply flick a switch in the cockpit and an emergency alarm would immediately flash to ground control. An ELT device emits a distress signal on three separate megahertz frequencies on the aircraft's band network. It was designed to do this in order to guarantee reception of the distress signal anywhere in the world.[42]

As of the date of this book's publication, none of Flight 370's three ELTs have been recovered and—if official reports are to be believed—no automatic emergency distress signals were picked up from any of the three ELTs known to be on board the aircraft at takeoff.

Black Boxes

Black boxes are those virtually indestructible information recorders that survive a plane crash. You may have heard the old adage about them, "Why don't they make the *whole plane* out of that stuff?" Too expensive, apparently.

Interestingly, they are not actually black in color. "Black box" is just an engineering term for a system that retains scientific information. They are actually orange.[43]

And *yet again* contrary to popular misconception, there is not *one* black box on a plane, there are two: the cockpit voice recorder (CVR) and the flight data recorder (FDR).

Cockpit Voice Recorder

The cockpit voice recorder is one of the black boxes and is a recorder that runs *automatically* from the moment a 777's engines are started. The cockpit voice recorder records all communications between crew and ATC, as well as all sounds and conversations that take place in the cockpit of the aircraft.[44]

The CVR continues running automatically and does not shut itself off until five minutes after the final engine shutdown.[45] *It cannot be manually shut off.* It does not even have an off switch. It does have an "erase" button, but the erase button will only function if the aircraft is on the ground with the parking brake set.[46]

However, the CVR records in a two-hour loop; in other words, it records over itself at the end of every two hours of recording. That would prove problematic if Flight 370 actually did fly for an extended period of time on autopilot.[47]

In addition to being made of virtually indestructible material, the CVR is actually mounted on the tail of the aircraft, rather than in the cockpit. The purpose of mounting it on the tail is to further minimize any possible damage to it during a crash.[48]

Most importantly, the CVR has an emergency transmitter that is automatically activated by impact or submersion in water. It also has an independent power source that will emit an emergency signal for thirty days. The system is designed to enable the location of high frequency sonar pings that are capable of being detected in deep waters.[49]

Yet, as of the date of this book's publication, the CVR has not been recovered and—if official reports are to be believed—no automatic emergency distress signals were ever picked up from it.

Flight Data Recorder

All technical information about the plane and its flight are automatically recorded by the flight data recorder; this means pilot control inputs, electronic status, control surface positions, speed, altitude, a record of all geographical positions, and a host of other information that is helpful in determining the cause of a problem.

Like the CVR, the FDR also has an emergency transmitter that is automatically activated by impact or submersion in water, and it too has an independent power source that will continue to emit an emergency signal for thirty days after. The system was also designed to enable the location of high frequency sonar pings that are capable of being detected in deep waters.[50]

Like the three emergency locator transmitters and the cockpit voice recorder, the flight data recorder—again, if official reports are to be believed—has not been recovered as of the date of this book's publication, and no automatic emergency distress signals were ever picked up from any of them.

That's an *extremely* unlikely occurrence, as one might well imagine.

Primary Flight Control Computers

Contrary, yet again, to many people's assumptions, there is not one, but a system of *nine computers* that run a Boeing 777. And "run it," they do. The model is so advanced that the coordinates for its planes, the altitudes at which they fly, their advance navigational waypoints, even possible alternate routes for their journeys, are information programmed into the flight directors' computers before the planes even take off.

The 777 has been in service since 1995 and its designers were highly cognizant of the fact that it is not humanly possible to envision *all* possible conditions that a plane may encounter in flight. Therefore, they intentionally designed a system that has different levels of safety features and security redundancies *built in*.

John Choisser explains this in *MH370: Lost in the Dark*. He writes, "The B777 has three flight control computers, each with three independent channels, for a total of nine computers, the majority of which must agree on their control outputs. How the software, hardware, and power supplies are isolated and integrated is a science in itself, and the results have provided extremely reliable service over the years."[51]

So, even if one computer fails, others are running. Those built-in redundancies have proven to be very accurate over a number of years of service in the 777 aircraft; even evoking the term "fail-safe." "The flight computer is programmed to handle the failure of various parts of the airplane," Choisser writes. "For example, in the case of a partial electrical failure, it must determine what components or functions are to be shut down, and in what order."[52]

The "multiple safeguards design" is also employed in the electrical systems of the plane by the electrical load management system (ELMS). The ELMS manages the aircraft's electrical loads and the protection of all of the aircraft's electrical systems to ensure that power is always available to the critical equipment that needs it. "The power system's reliability is very high," writes Choisser, "and there are numerous ways power can be switched between busses, depending upon the type of failure and what power is available from which sources."[53]

Even in the event of total electrical failure, the systems on board the aircraft would be able to continue to operate and it could continue to fly.[54] In other words, the notion that one problem or one decision from someone could knock the plane right out of the sky is about as farfetched a scenario as anyone can come up with.

The programming of the flight control computers is done by inputs from the pilots, the automatic pilot system, the autoland control, and the flight director itself. This makes for a very complex system where a computer, the flight director, programs and controls other systems, and the flight control computers control the functions of the aircraft during flight.

Fly-by-Wire System

Fly-by-wire is a relatively new system that actually eliminates the direct connections between the pilot and the aircraft controls. It does so by using electrical inputs that bypass the flight crew and go directly into the flight control computer. The Boeing website explains this as follows:

> The flight-control system for the 777 airplane is different from those on other Boeing airplane designs ... Boeing designed the 777 with fly-by-wire technology. As a result, the 777 uses wires to carry electrical signals from the pilot control wheel, column, and pedals to a primary flight computer.[55]

The engineers of the 777 could even have opted, if they'd wanted to, to give the aircraft computer systems total control over the pilots. It's that capable a feat of engineering. For example, in an Airbus 320 plane, the flight computers have so much control that some consider the role of the pilots of the plane as though they are just playing a simulated game. But on the Boeing 777, those in charge of the engineering decisions opted to at least give the pilots the *ability* to override the flight control computer systems, if they choose to, in the event of an emergency. It is no exaggeration to say that flight computers can fly a plane without the pilots. They have that capability. But the 777 incorporates inputs from the flight crew to avoid the potential dangers of *too* automated a flight system. For example, there was a case in Eastern France in 1988 where the 320 crashed because it came in so low that the Airbus computer took over and automatically attempted to land the plane because of the altitude reading and it crashed into the trees in the process.[56] That's a vivid example of too much technology. To avoid this, in the 777, the pilots oversee the various processes during flight, and the landing is the result of inputs by the flight crew. That is why pilots rightfully resist the notion that they are now just babysitters for the automatic flight systems of the plane; they *are still* a very necessary function of safe air travel. As Patrick Smith puts it:

> The autopilot is a tool, along with many other tools available to the crew. You still need to tell it what to do, how to do it, and when to do it. I prefer the term *autoflight system.* It's a collection of several different functions controlling speed, thrust, and both horizontal and vertical navigation—together or separately, and all of it requiring regular crew inputs to work properly.[57]

If you're anything like we were with wrapping your brain around all this information, you are starting to see that a flight is not only determined by what the pilot decides to do in the cockpit of the airplane. The 777 is a highly sophisticated aircraft that can, itself, and *does*, itself, operate the flight functions of the plane, even prioritizing systems when problems arise, and always determining the safest and most secure manner in which to proceed. The survivability of the aircraft is built in by design. As of March 8, 2014, the day 370 went missing, the survivability of the 777 had been tested—and proven—over a period of nineteen years.

Automatic Fire and Smoke Suppression Systems

Smoke detectors and fire suppression systems are located throughout a 777 aircraft. They utilize the chemical solution Halon 1301, a fire suppressor. Each of the plane's three cargo departments is equipped with smoke detection systems.[58] There is a system in place to automatically address smoke and fire.

According to Choisser, "When fire is detected, two bottles of Halon are discharged immediately in a "knockdown" dose, followed by a metered dose over the next hour or so from the remaining three bottles. Fire warning in the cockpit alerts the crew to a cargo fire and its location, after which the pilot arms the appropriate cargo area and pushes the discharge button. This also automatically causes the air circulation and air conditioning/heating systems to modify their settings."[59]

There are a variety of ways a fire could spread on an aircraft, and one is from the presence of highly flammable materials—lithium-ion batteries, for example. We call out this material specifically because there was reportedly a large shipment of lithium-ion batteries on board Flight 370, in one of its cargo departments. Lithium-ion batteries are known for their ability to explode and burst into flames, and they were a key factor in the crash of a plane four years before Flight 370.

Choisser writes, "In September of 2010, a B747 UPS cargo flight with thousands of lithium batteries on board suffered a cargo fire and crashed after leaving Dubai. Smoke in the cockpit was a serious problem in the unsuccessful effort to … [land safely in] Dubai.[60] …This event resulted in an improvement in cargo regulations and a restriction on passenger flights carrying lithium-ion batteries."[61]

Upon review, it is not clear whether or not the fire suppression systems of a 777 are sufficient in the case of a lithium-ion originating fire. The reason is that the best way to put out a lithium-ion battery fire is with water, and although flight attendants are specifically trained to put out battery fires with water in the case of personal electronic devices in the passenger cabin, if a fire occurs in the cargo bay of the aircraft, then Halon, instead of water, will be dispensed.[62]

Cockpit Security

The door to the cockpit of a 777 meets security standards for ballistics because it's bulletproof and intruder resistant. It locks automatically, electrically, and has a manual lock as well. Choisser describes it, "The door opens inward toward the pilots. When it is closed and electrical power is available, the door locks. It unlocks when electrical power is removed. There is a viewing lens in the door so that pilots can see into the passenger cabin. The pilots can open the door by turning the knob … There is also a deadbolt lock with a key for access. The lock can be set manually from the flight deck side of the door to allow or not allow access with the cabin key."[63]

There are also elaborate security protocols for access to the cockpit. "The emergency access system allows access in case of pilot incapacitation. There is an electronic code pad on the passenger cabin side of the door. When the emergency access code is entered, there is a time delay before the door opens. During this delay, the pilot can deny entry if he is not incapacitated. In the event of rapid decompression, the door unlocks."[64]

Electronics Equipment Bay

The electronics equipment bay, or E/E bay as it's known, contains all the electrical systems of an aircraft and is located underneath the cockpit. It's a pretty large room with a lot of equipment. It also contains the pilots' oxygen supply. If you'd like to see what it looks like, simply search online for "777 E/E bay" or enter the URL www.youtube.com/watch?v=s3suAKXZ3BQ.

Since the location of the E/E bay is underneath the cockpit, it is generally considered secure. However, we unearthed some disturbing evidence that reveals that it's not as secure as one would hope. The E/E bay in a triple-seven can apparently also be accessed from a latch on the floor of the passenger cabin that is surprisingly unsecure. In fact, Australian pilot Captain Matthew Wuillemin placed a video of it on YouTube in the hopes that the problem would be corrected.[65] You can see the problem pretty vividly if you search online for "B777 E/E ACCESS" or go to www.youtube.com/watch?v=mLmzvF2qkDY.

In the text featured with the video, Capital Wuillemin has written, "The following video shows that access can be gained from sensitive areas of an aircraft from the passenger cabin while in flight. So after all the changes since 9/11, with screening and restricted items being on board, this would seem to be something that requires attention to prevent a repeat of those events. Around 3000 aircraft are estimated to be affected. The fix is simple, so why not apply it?"[66]

Since the E/E bay contains a plane's circuit breaker panels, as well as the flight crew's oxygen, it's pretty disturbing to think that it could easily be tampered with. According to Captain Wuillemin, the hatch is "NOT locked or secured [making the] B777 Aircraft vulnerable to incursion."[67]

Emergency Oxygen Supply

The emergency oxygen supply system in the cockpit is more sophisticated than that of the passenger cabin. It is designed to get the pilots through a decompression or a fire so that they can continue to control the plane.

Choisser writes, "The pilots' oxygen masks have an inflatable head harness for a quick and snug fit. The pilots' masks prevent breathing of smoke or noxious gasses, and must be set at 100% oxygen if smoke is present. The amount of oxygen must allow at least 12 minutes of oxygen, which is enough time to get the airplane down to 10,000 feet on most routes. Some airlines in South America and Asia have mountainous terrain that requires an increase in the time to 22, or in some cases involving the Himalayas, 70 minutes,"[68] In other words, if there is a loss of oxygen on a plane, the pilots will quickly move the plane to lower altitudes and at those lower levels there is sufficient oxygen to breathe.

In addition to supplying the necessary oxygen levels and featuring a smoke suppression feature, the pilots' oxygen masks contain a microphone for communication purposes and the microphone activates automatically when the oxygen mask is put on.

Emergency Doors

Another thing that's a lot safer in the aircraft than you might think are those emergency exit doors. There seems to be a ridiculous, but highly prevalent notion that all that has to happen for a plane to implode is for one person to be crazy enough to open one of those door latches while the plane is airborne— and then, the theory goes, everybody will get sucked out of the plane like fleas into a vacuum cleaner the size of ten galaxies. NOT TRUE, boys and girls.

Again, the misunderstanding is a case of asking the wrong question. The question shouldn't be, "What would happen if someone opened a door while the plane was airborne?" It should be, "Can you actually open those doors during flight?" And the answer to that question is no.

Patrick Smith writes:

> While the news never fails to report these events, it seldom mentions the most important fact: You cannot—repeat, cannot—open the doors or emergency hatches of an airplane in flight. You can't open them for the simple reason that cabin pressure won't allow it. Think of an aircraft door as a drain plug, fixed in place by the interior pressure. Almost all aircraft exits open inward. Some retract upward into the ceiling; others swing outward; but they open inward *first*, and not even the most musclebound human will overcome the force holding them shut. At a typical cruising altitude, up to eight pounds of pressure are pushing against every square inch of interior fuselage. That's over 1,100 pounds against each square foot of door. Even at low altitudes, where cabin pressure levels are much less, a meager 2 p.s.i. differential is still more than anyone can displace—even

after six cups of coffee and the aggravation that comes with sitting behind a shrieking baby. The doors are further held secure by a series of electrical and/or mechanical latches.[69]

Bullet-Proof Elements

Also check out the truth about a bullet fired on a plane. You've probably seen a lot of movies where a bullet gets fired while airborne and every man, woman, and pretzel on that plane immediately gets sucked out of that hole caused by the bullet as a result of the rapid decompression. *Not possible.*

Marshall Brain, the founder of HowStuffWorks.com, explains: "If the bullet simply punctures the skin of an airplane, then it's no big deal. The cabin of the airplane is pressurized, and the hole creates a small leak, but the pressurization system will compensate for it. A single hole, or even a few holes like this, will have no effect."[70]

In an episode of the TV show *Mythbusters*, after the hosts fired a bullet from a nine millimeter Glock at the hull of a plane in a fully pressurized cabin, here's what happened, as documented by Deane Barker on gadgetopia.com:

"Nothing. Sure, the air rushed out, but even Styrofoam peanuts they had placed in the aisle didn't move, much less Buster [the crash dummy strapped into a chair]. They did the test a second time, this time firing through the window right next to Buster. Same result—nothing."[71]

You can see it for yourself at www.youtube.com/watch?v=Fi1_1l7M8FA.

After firing the bullet failed to provide anything spectacular, they set off a small explosion against the hull of the plane; "they rigged explosive cord around the window to simulate the window blowing out due to structural failure. They pressurized the plane again and blew the window.[72] The results [this time] were much more satisfying—Buster got his arm yanked out the window. Crash test dummies aren't designed to come apart, but if he were a human, I think he would have lost his arm. However, the rest of his body just wasn't going through the window, no matter what the myth said."[73]

Blowing out an entire window, i.e. a very large hole, would apparently cause the type of emergency from rapid decompression that those films depict. However, it would require a massive explosion to do that, not just a single bullet, and getting the equipment on board that's necessary to make this happen would be practically impossible since security almost tries to shut down an airport these days if they so much as spot somebody trying to take a bottle of water on a plane.

If you start paying less attention to the confounded claims spouted by mainstream media and paying *more* attention to what actual aircraft experts and accident investigators say, you will notice, as we did, a common thread that keeps reoccurring. Goglia sums it up:

"For all communication to suddenly cease without a distress signal usually indicates a catastrophic failure of the aircraft, not allowing time for the crew to communicate either by radio or through the aircraft transponder. Modern airliners have multiple radios for voice communication and the transponder can be used to send signals that indicate different problems with the aircraft (for example a discreet code for hijacking)."[74]

So that's your "what-to-take-away-from-it-all" message. The Boeing 777 is such a sophisticated plane that there are simply too many built-in fail-safe systems for them to, under almost any circumstance, *all* fail at once. There would have to be a catastrophic event for this to be remotely plausible.

Two

Malaysia Airlines

Before 2014, Malaysia Airlines was often described as having an excellent record of service and safety.[75] In fact, its airline safety rating was higher than most airlines, based on the airline safety ratings website AirlineRatings.com, www.airlineratings.com/safety_rating_per_airline.php. Other reports, however, suggest a record of several near-misses and emergency landings, also several crashes.[76] However, most of those incidents occurred in the last century and with other types of aircrafts than the 777.[77] The only major safety incident involving a Malaysia Airlines Boeing 777 was on August 1, 2005.[78] An article by the *Mirror* reads:

"Boeing 777-2H6ER shoots to 3,000 feet shortly after takeoff in Perth, Australia after instrumentation problem. Pilot forced to take manual control and make emergency landing. No fatalities."[79]

Other than the instrumentation problem in 2005, there were only two recorded problems of a serious nature with the model between the time it entered service in 1995, and before Flight 370 in 2014. According to the *Mirror Online*:

> The first was on January 17, 2008, when a BA [British Airways] flight from Beijing to Heathrow crash-landed after developing a fuel system fault, causing 45 injuries, one of which was serious. The fuel system on this model was subsequently re-designed.
>
> The second incident took place on July 6, 2013, when an Asiana Airlines flight from South Korea crashed into a sea wall as it approached San Francisco International Airport, causing three deaths and 180 injuries. Pilot error is understood to have been the cause of the crash.[80]

Bear in mind that these incidents, although serious, are the *worst* problems that have been noted through nearly *five million flights* of Boeing 777 planes.[81] So, all things considered, prior to Flight 370, Malaysia Airlines was recognized as a relatively safe air carrier, and the Boeing 777 was universally regarded as one of the safest, most modern airplanes in service.

Three

Events of March 8, 2014

You can listen on YouTube to the voice transmissions that took place between the Malaysia Flight 370 flight crew and Kuala Lumpur air traffic control. The words are haunting and mundane at the exact same time, in that there is absolutely nothing unusual in their language, yet we know they are among their last words before something very dramatic happened.

Some articles have attempted to make something mysterious out of the final radio contact received from the flight, because the last words were "Good night, Malaysian Three Seven Zero." However, if you examine the recording, as we did, you will note from those transmissions that there is nothing at all out of the ordinary. "Good night, Malaysian Three Seven Zero" was simply the final transmission, a standard sign-off from the cockpit of the plane, telling air traffic control in Kuala

Lumpur that MH 370 copied the last transmission from ATC
and was leaving their airspace control en route to Vietnamese
airspace. That's known as the "handing-off" process—in this
case, from Kuala air traffic control to Ho Chi Minh air traffic control—and it's standard operating procedure. That final
transmission took place at 1:19 A.M., to be precise, 1:19 A.M.
and 29 seconds. Five seconds earlier, at 1:19:24 A.M., ATC-
Kuala Lumpur bid goodbye to MH-370, simply due to the fact
that the flight was leaving their control area. They directed the
crew to contact ATC-Vietnam, normally the next control area
to pick up the flight. The exact transmission was, "Malaysian
Three Seven Zero, contact Ho Chi Minh one two zero decimal
nine. Good night."

We know, therefore, by recorded transmissions, that, as
of 1:19:24 A.M., all was routine with the flight of Malaysian
Three Seven Zero. For anyone not completely familiar with
the events of that auspicious day, we compiled a much more
detailed timeline than any we have found in our research on
this matter:

TIMELINE: FLIGHT 370

Saturday, March 8, 2014

Shortly after midnight:

227 passengers and twelve crew members begin boarding Malaysia Airlines Flight 370. 154 of the passengers are
Chinese/Taiwanese; thirty-eight passengers are Malaysian; the
rest are from India, Indonesia, Australia, France, New Zealand,
Ukraine, Canada, Russia, Italy, Austria, and the Netherlands,
and three passengers are from the United States. The crew
members are Malaysian. The pilot, Captain Zaharie Ahmad

Shah, is a fifty-three-year-old veteran with 18,365 flying hours. The captain joined Malaysia Airlines in 1981. The first officer (also referred to as the copilot) is Fariq Ab Hamid, who has 2,763 flying hours. Hamid is twenty-seven years old and has been with the airline since 2007.

12:40 A.M.

Flight 370 is prepared to depart Kuala Lumpur International Airport. The aircraft is a Boeing 777-200ER, a plane with an excellent safety record. Malaysia Airlines has fifteen of the 777-200 ERs in its fleet. It is on its way to Beijing, a trip of approximately twenty-seven hundred miles. It is scheduled to land there at 6:30 A.M. The weather is clear and, concurrently, the quarter moon is setting in the west.

12:40:38 A.M.

Flight 370 is formally cleared for takeoff on runway 32R. The pilot confirms takeoff on 32R and begins his takeoff roll.

12:41 A.M.

Flight 370 takes off.

12:42:05 A.M.

Pilot reports that they are airborne and leaving the airport area.

12:42:10 A.M.

Kuala Lumpur air traffic control confirms and clears Malaysia 370 to rise to flight level 180 (eighteen hundred feet above ground) and directs it to turn right, toward IGARI waypoint (the navigational direction for its northeast flight path).

12:42:40 A.M.

Pilot confirms receipt of course and altitude instructions from KL Airport.

12:42:52 A.M.

KL Airport hands off Flight 370 to Kuala Lumpur radar on frequency 132.6, and Flight 370's pilot confirms hand-off.

12:46:51 A.M.

Pilot contacts Kuala Lumpur air traffic control. KL-ATC instructs 370 to climb to flight level 250.

12:46:54 A.M.

Pilot confirms 370 is climbing to level 250.

12:50:06 A.M.

KL-ATC directs 370 to climb to level 350.

12:50:09 A.M.

Pilot confirms 370 is climbing to level 350.

01:01:14 A.M.

Pilot informs KL-ATC that 370 has reached and is remaining at level 350.

01:01:19 A.M.

KL-ATC confirms level.

01:07:55 A.M.

Pilot again informs KL-ATC that 370 is remaining at level 350.

01:08:00 A.M.

KL-ATC again confirms level.

01:19:24 A.M.

KL-ATC hands off 370 to Vietnam air traffic control at frequency 120.9, saying: "Malaysian Three Seven Zero, contact Ho Chi Minh one two zero decimal nine. Good night."

01:19:29 A.M.

370 pilot confirms hand-off, saying "Good night, Malaysian Three Seven Zero."

01:21 A.M.

Transponder makes last automatic contact from Flight 370, near IGARI waypoint (approaching southern Vietnam on a north-northeast heading).

01:30 A.M.

Forty-nine minutes after takeoff, Vietnam air traffic control loses contact with the plane over the sea between Malaysia and Vietnam, at the following coordinates: 06 55 15n 103 34 43e.

This is when the aircraft's transponder was either turned off or stopped working.

Vietnam ATC locates a flight near those coordinates, Flight MH52, bound for Osaka and about thirty minutes ahead of Flight 370, and requests the plane attempt to radio contact 370. MH52 reports back to Vietnam ATC that their radio contact attempt *was* answered, but the response was garbled and unintelligible.

Vietnam ATC reports to Malaysian authorities that Flight 370 has turned back toward the west, but receives no response. Vietnam ATC reports a second time to Malaysian authorities that Flight 370 has turned back toward the west, but again receives no response.[82] It is reported that at the time the aircraft transponder stopped working, the plane had approximately 7.5 hours of fuel remaining.

01:37 A.M.

The automatic ACARS transmission due from Flight 370 at 01:37 A.M. does not arrive.

One needs to look not only at what happened, but also at what did *not* happen.

At this point—and probably earlier—it has to be clear to all concerned that something is dramatically wrong. Typically, top priority is given to a missing flight—all forms of emergency communications are immediately enabled and military fighter jets are quickly sent to locate the missing plane and escort it to safety.

Something either precluded these events from taking place or they have not been reported.

The following statement from the Prime Minister of Malaysia, Najib Razak, on March 15, 2014, is an important addendum to the known facts of the Flight 370 timeline:

> Based on new satellite information, we can say with a high degree of certainty that the aircraft communications addressing and reporting system (ACARS) was disabled just before the aircraft reached the east coast of peninsular Malaysia. Shortly afterwards near the border between Malaysian and Vietnamese air traffic control, the aircraft's transponder was switched off.
>
> From this point onwards, the Royal Malaysian Air Force primary radar showed that an aircraft which was believed—but not confirmed—to be MH370 did indeed turn back. It then flew in a westerly direction back over peninsular Malaysia before turning northwest. Up until the point at which it left military primary radar coverage, *these movements are consistent with deliberate action by someone on the plane.*

Today, based on raw satellite data that was obtained from the satellite data service provider, we can confirm that the aircraft shown in the primary radar data was flight MH370. After much forensic work and deliberation, the FAA, NTSB [National Transportation Safety Board], AAIB [Air Accidents Investigation Branch], and the Malaysian authorities, working separately on the same data, concur.

According to the new data, the last confirmed communication between the plane and the satellite was at 8:11 A.M. Malaysian time on Saturday 8 March. The investigations team is making further calculations which will indicate how far the aircraft may have flown after this last point of contact. This will help us to refine the search.

Due to the type of satellite data, we are unable to confirm the precise location of the plane when it last made contact with the satellite. However, based on this new data, the aviation authorities of Malaysia and their international counterparts have determined that the plane's last communication with the satellite was in one of two possible corridors: a northern corridor stretching approximately from the border of Kazakhstan and Turkmenistan to northern Thailand, or a southern corridor stretching approximately from Indonesia to the southern Indian Ocean. The investigation team is working to further refine the information.

In view of this latest development the Malaysian authorities have *refocused their investigation into the crew and passengers on board.* Despite media reports that the plane was hijacked, I wish to be very clear: we are still investigating all possibilities as to what caused MH370 to deviate from its original flight path.[83]

The timeline points were primarily constructed from information in Choisser's *MH370: Lost in the Dark*,[84] Bloomberg Visual Data,[85] and CNN.com.[86]

Four

"Official Conclusion," The Plane Crashed into the Indian Ocean

Malaysia Flight 370 left Kuala Lumpur Airport shortly after midnight on March 8, 2014. Everything was normal—all communications were routine—until, at 1:20 A.M., it lost all contact with *everything, everyone, and in every way*. Flight 370 "disappeared on 8 March, 2014, at 01:20 after losing contact with air traffic control less than an hour after takeoff."

Finally, more than two days later, on March 10, 2014, an official press conference delivered a more "official explanation" (for lack of a better term), although it was empty of anything close to adequate. They basically reiterated that they were still searching for the plane and had not ruled anything out. After they read the official statement, the questions were rather blunt:

Question One: Is DCA [the Department of Civil Aviation] narrowing down the investigation as [a] possible mid-air explosion or mid-air disintegration?

Answer: At this stage, we did not narrow possibilities. We are looking at all angles of what could possibly have happened on that ill-fated flight.

Question Two: Usually when a plane disappears or goes down, there will be a transmitting signal from the black box and it should continue to pump out the signal. Did you get any signal right now from under the sea?

Answer: From the day the aircraft lost contact from radar, there is no signals [that] been able to be detected by any of our ships or aircrafts in the region.

Follow-Up: Any reason why there is no signal?

Answer: If we know the reason we would tell you.

Question Three: The beacon, is it a pilot-automated or does it goes off on its own?

Answer: It goes off on its own, on impact and when they touch water.

Follow-Up: On the rescue operation, will you double check the areas that have already been covered? Now it's fifty nautical miles.

Answer: Yes, we will double check. We will direct our aircrafts in such a way that they fly so that they do not miss anything. So as for the ship.

Question Four: Did the stolen passport holder have Asian features? Who has jurisdiction over the disappearance of the missing aircraft?

Answer: I would not reveal too much on the two passengers because they are under investigation [and] because any information that I give you might jeopardize [the] investigation that is going on; but on the areas of us searching covers both Malaysian waters and Vietnamese waters.

Question Five: Have the technical assistance teams from the NTSB, Boeing, the FAA, and the French accident investigators, which I believe have offered their assistance, have they arrived and have they started to do their work?

Answer: The offers that we got are from FAA, NTSB USA, AAIB UK, and also the Australian, we have not got anything from [the] French yet. The FAA and NTSB USA, I was informed this morning, they are here and we are going to meet them as soon as possible to discuss with them how could they offer their assistance to our investigations of this incident.

Question Six: Are we investigating a group from China who's currently claiming responsibilities?

Answer: Can you please give more information because we have not received anything.

What has been reported is something that we need to verify officially. That is more on security matters.[87]

Question six was in reference to an early report that a small terrorist group in China claimed responsibility for the act. The group is known as the "Chinese Martyrs Brigade," which is reportedly a sect of China's Muslim community called the Uighurs, which has had ethnic tensions in China and is said to "have links with Al-Qaeda-affiliated group,

Jemaah Islamiyah, which has a presence in Malaysia, as well as the Philippines and Indonesia."[88]

That group reportedly claimed responsibility for what happened to the missing Flight 370, with a statement indicating that the reason for the act was to murder the largely Chinese passengers on board the plane as retribution:

"You kill one of our clan, we will kill 100 of you as payback."[89]

However, the claim was difficult to verify—it was posted on an encrypted anonymous message board called "Hushmail" on March 9, 2014—and the director of the CIA, John Brennan, seemed to attach no importance to the claim. On March 12, 2014, he, as described by the global news network RT, rt.com, "said that there have not been any credible claims of responsibility from terrorist groups for the plane's disappearance."[90] He did, however, make some cryptic remarks about terrorism, saying that the CIA had "*not at all*" ruled out terrorism as possibly having played a part in the disappearance of Flight 370."[91] As director of the Central Intelligence Agency—we assume, hope, that he was aware of the claim by the Chinese Martyrs Brigade at the time that he made that statement.

Of course, the CIA officially still maintains that President Kennedy's complex and multi-layered assassination was pulled off by an ex-marine with a rifle that couldn't even shoot straight, so what the hell do they know?[92] If anything, maybe the way that the claim was completely ignored should encourage us to place *more* focus on the report rather than less, as was seemingly intended. The claim certainly didn't sound inconsequential or harmless.

It was later announced that the flight also made two course changes. We should point out that there's no surefire way for us to verify it, but this is what the public was told by authorities. Remember, as you consider those changes, that they would have to have been preprogrammed into the flight computer during the early portion of the flight.

Greg Feith, a former investigator for the National Transportation Safety Board, told NBC News, "The flight path change was first confirmed by authorities on Saturday [March 15], who said the aircraft had veered off course because of 'deliberate action by someone on the plane.'"[93]

The course change was an extreme departure from the official flight path, which was north-northeast, to a western flight path that rerouted the plane back over the Strait of Malacca, toward the Indian Ocean. It later changed course yet again, turning south over the Indian Ocean.[94]

Abrupt altitudinal changes were also reported. Military radar indications say the plane climbed to a level of forty-five thousand feet and turned sharply west. It then reportedly descended to twenty-three thousand feet, before again ascending to a flight level of thirty-five thousand feet.[95]

About ten minutes later, the jet is "thought to have dropped" to five thousand feet, as Malaysia military radar continued to reportedly monitor the aircraft over the Strait of Malacca, wildly off its course.

The Australian Safety Transport Bureau posed that oxygen loss was a likely a factor on Flight 370 and may have caused the death of all on board as the result of hypoxia.

Australian authorities said they believe that someone on board Malaysia Airlines Flight 370 switched on the autopilot system deliberately after the plane turned toward the southern Indian Ocean. They also theorized that all 239 passengers and crew had become unresponsive, possibly after being deprived of oxygen, before the plane ran out of fuel and crashed.

Those were the main reasons the Australian Transport Safety Bureau gave in a report Thursday for setting a massive new search area—the third in as many months for the airliner, which disappeared March 8. The new hunt is slated to restart in August as much as 600 miles south of the previous underwater focus.

Australian Deputy Prime Minister Warren Truss said it was "highly, highly likely" the autopilot was switched on deliberately after the plane had veered off its assigned course from Kuala Lumpur to Beijing.

That is more definitive than investigators have been about human intervention setting up the flight path toward one of the most remote sections of the globe.[96]

Many people, particularly members of the media, have asserted—incorrectly—that oxygen deprivation would have caused a horrible death by suffocation for everyone on the aircraft. That's medically incorrect, and in a big way. Oxygen deprivation causes hypoxia. And for what it's worth, hypoxia from oxygen deprivation as a result of rapid decompression is supposedly a very painless death.[97] Some people have recommended it as a more "humane" method of capital punishment, which, we have to point out, is inherently *in*humane. Regardless, pilots have been taught how to react and what emergency procedures to employ if a plane's flight monitoring systems warn that there's not enough oxygen in it.

After many more months, here is the "official" conclusion regarding the missing Boeing 777 and the 239 human beings who were on board: "Investigators say what little evidence they have to work with suggests the airplane was deliberately diverted thousands of kilometers before eventually crashing into the Indian Ocean off the coast of Western Australia."[98]

As far as the plane and what happened to it: *STILL looking, folks. We'll let you know if anyone finds anything.*

Five

Multiple Eyewitness Testimonies You Probably Haven't Heard About

Police investigators are rightfully cautious in using eyewitness testimonies because they tend to vary widely. When they *don't vary*, one would think there is reason to allot more importance to them. If four eyewitnesses in four different locations all said they saw a jumbo jet that matched the description of the missing airplane, then shouldn't investigators be carefully considering when they said it happened and where they said they were when they saw it? You better believe it, boys and girls. And there were actually a lot *more* than four eyewitnesses. So, did the mainstream media assess importance to the eyewitness testimonies in this case? Oh *hell no*. They basically *buried* them! They never connected the dots, never matched up the various eyewitness reports to see what patterns they form. They mentioned the eyewitnesses only briefly near the time of the original incident, and then never followed up on the significance of their reports.

Let's back up a bit. It should be noted, first and foremost, that the *geographical locations of the eyewitnesses' sightings are consistent with the reported course change that redirected* Flight 370 from its planned northeast route toward Beijing via Vietnam, into a heading in which it veered dramatically off course, heading southwest toward the middle of the Indian Ocean. Coincidentally or otherwise, a major US air base, Diego Garcia, also happens to be in the middle of the Indian Ocean, and that's one of the theories we will be exploring later in this book.

Unlike the mainstream media, we will show you those eye-witness reports through a time window that makes sense of their geographical locations. There were at least five locations of multiple eyewitness sightings of the plane that appear to be quite viable.

First, let's look at the testimonies that align with the time the plane lost contact with air traffic control. There are at least *thirteen eyewitnesses* who said, with certainty, that they saw a large commercial airliner that was either in flames or flying unusually low. Four of those testimonies noted locations and times that were right around the location and time that Flight 370 lost contact with air traffic control. They are:

- An oil rig worker off the coast of southern Vietnam.[99]
- A businessman and two fishermen in Kota Bharu, Malaysia, who all reported that they saw the plane's lights, as it was flying surprisingly low.[100]
- A bus driver in Penarik, Malaysia, southeast of Kota Bharu, along the coast of Malaysia, who was also surprised to see a very low-flying jetliner.
- Eight villagers near Marang, Malaysia, southeast of Penarik and again along the coast of Malaysia, who told police that they "heard a loud and frightening noise which sounded like the fan of a jet engine."

Those thirteen witnesses were all in a location at a time that would concur with when Flight 370 supposedly went off course. Most of their locations can be seen on a map at the following web address, www.dailymail.co.uk/news/article-2578914/ Nine-fresh-witnesses-place-missing-jet-near-Thailand-despite-Malaysia-military-moving-search-area-west.html.

Let's consider them one by one.

EYEWITNESS SIGHTINGS BY LOCATION

Sona-Mercur Oil Rig, Southeast of Vũng Tàu, Vietnam

Mike McKay is an oil rig worker from New Zealand and on March 8, 2014, was stationed at an oil rig off the coast of Vietnam. The rig is called the Sona-Mercur oil rig and its location is approximately 186 miles southeast of a coastal Vietnam city named Vũng Tàu. The location lines up with the intended flight route of Malaysia 370 from Kuala Lumpur, Malaysia, to Beijing, China. Look it up online and you'll see what we mean.

The timing of McKay's sighting was also in line with the "disappearance" of Flight 370, a few hours after the plane had taken off at shortly after midnight. McKay was certain he witnessed a burning plane, which he believes was Flight 370, above him in the early morning hours. In fact, his certainty was sufficiently strong that, sensing its importance, he documented each fact he recalled in a detailed message that he then sent out to his company, in the expectation that it would be passed to the appropriate sources and help them with the recovery of Flight 370. When he was *not contacted* by any of the appropriate authorities in response to his message, he sent the following email to news sources in the hope that they would follow up on his information. The following is a verbatim copy of that email; we have not changed anything, except that his passport number is not shown for obvious reasons, so all emphasis and grammar you see was in the original email:

Gentlemen,

I believe I saw the Malaysian Airlines plane come down. The timing is right. I tried to contact the Malaysian and Vietnam officials several days ago. But I do not know if the message has been received.

I am on the oil-rig "Songa Mercur" off Vung Tau.

The surface location of the observation is:

Lat. 08° 22' 30.23" N

Long. 108° 42' 22.26" E.

I observed (the plane?) burning at high altitude and on a **compass bearing of 265° to 275°** from our surface location.

It is very difficult to judge the distance but I'd say 50–70 km along the compass bearing 260°–275°.

While I observed the burning (plane) it appeared to be in <u>ONE piece</u>.

The **surface sea current** at our location is: 2.0–2.3 knots in a direction of 225°–230°.

The **wind direction** has been NE–ENE averaging 15–20 knots.

From when I first saw the burning (plane) until the flames went out (still at high altitude) was 10–15 seconds. There was **no lateral movement**, so it was either coming toward our location

Michael Jerome McKay

NZ passport #[101]

It should be noted that the email was verified by McKay's employer,[102] but McKay was reportedly harassed and even lost his job as a result of trying to do the right thing. Sy Gunson posted about this on Reddit, writing:

"Mike McKay was under pressure from employers not to talk to the media or the public and has stopped using the email address previously publicized. He only communicates to a handful of people. In May he wrote to a friend of mine:

'I do know the exact time of my observation (+/– 30 minutes).'

"He also pointed out to my friend that he made his original notification without the benefit of knowing what else had been reported. The New Zealand Police were provided a full statement.

"He has since lost his job due to the displeasure of his employer with his efforts to notify his sightings to authorities. He has not contacted the media. They have attempted to contact him.

Mike McKay has not benefited financially and has in fact lost his job because he stood on a matter of principle and tried to help."[103]

To summarize: McKay was sure of what he saw and gave the precise location he was in when he saw it, and the time and location line up with the time the flight signed off and said good night to Kuala Lumpur air traffic control and was expected to be contacting Vietnam air traffic control at Ho Chi Minh ATC—contact that was never made.

Also note that the sky was reportedly lit by at least a partial moon. By one report, "The moon was half-full at the time, making it a little easier to see."[104] By another report, it was a "quarter moon."[105] If McKay was looking up at a moonlit sky, he would have had no problem whatsoever seeing a burning plane.

Further credence to McKay's reported sighting is substantiated by this information, sourced by the *International Business Times*:

> On Tuesday [four days after the flight went missing], a Malaysian newspaper quoted the commanding general of the nation's air force as saying that the plane had been tracked by military radar up to two hours after takeoff, until a point well to the west of where civilian authorities had said it was. That point was over the Straits of Malacca, on the way to the Indian Ocean, indicating the jet had turned back and was flying due west, not north-northeast toward Beijing. A military source confirmed the statement to *Reuters*.
>
> Then, later Tuesday, the general retracted his words, saying that the journalist had misunderstood, and the air force could not tell for sure that it had really tracked the missing airplane.

And one day later, the email from the rig worker surfaced.[106]

So quite a lot of obvious flip-flopping and obfuscation was going on. But that "turn back" would have put the jet right where McKay said he saw it.

Kota Bharu, East Coast of Malaysia

The night Flight 370 went missing, two fishermen had taken people fishing off the coast of Kota Bharu and were returning home when they were very surprised to see a very low-flying plane over the South China Sea at about 1:30 A.M. Azid Ibrahim, one of the fishermen, described it: "It was really low. I saw the lights. They looked like the size of a coconut."[107]

Ibrahim was familiar with the normal flight patterns of commercial jetliners in that area. He said, "I was fishing when I saw the plane—it looked strange. Flying low. I told my friend that's not normal. Normally, it flies at 35,000 feet. But that night it touched the clouds. I thought the pilot must be crazy."[108]

A businessman, in a different area of Kota Bharu at the same time, reported that he saw an aircraft's lights low in the sky, and then it disappeared.[109]

Penarik, East Coast of Malaysia

According to Malaysian deputy police commander Dak Jalaluddin Abdul Rahman, "A bus driver, who gave his voluntary statement on Sunday [a day after the crash], said he saw a low-flying plane at Penarik at about 1:45 A.M. the same day flight MH370 went missing. The driver was sure that he saw the aircraft's blinking beacon lights."[110]

Marang Area, East Coast of Malaysia

Eight men, residents of Kampung Pantai Seberang Marang, in the Marang area of Malaysia, heard what's been described as an "explosion" at 1:20 A.M. on the day Flight 370 went missing. They said they were seated on a bench approximately four hundred meters from the Marang beach when they all heard the mysterious sound, "… which sounded like the fan of a loud jet engine." The men reported it to police later that day, saying they believed it was linked to the missing jetliner. Police in Terengganu confirm the report.[111]

West, Toward Indian Ocean

Further west, in line with the altered flight route taking the plane toward the Indian Ocean, there are also multiple eyewitness reports of a plane fitting the exact description of a Malaysia Airlines Boeing 777.

Maldives, West of Malaysia and South of India

On the remote island of Kudahuvadhoo, a local newspaper reported that multiple eyewitnesses described an occurrence they'd never seen before, an airplane near their island. They said what they'd seen was a low-flying jumbo jet—and described it as a plane that had the same markings as Flight 370. They said it was very low and very loud. The sighting occurred at a time and in a location that was *within the fuel range of Flight 370.*[112]

The witnesses stated that the jumbo jet was flying toward the southern tip of the Maldives. Coincidentally—or otherwise—that would have positioned it to be going in the general direction of the US air base Diego Garcia, though not by the most direct route from the point where Flight 370 changed course. A landing at Diego Garcia has been a theory on what might have happened to the plane.

According to *Haveeru Daily*, the leading news website in the Maldives, the witnesses noted the "incredibly loud noise that the flight made when it flew over the island."[113] One said, "I've never seen a jet flying so low over our island before. We've seen seaplanes, but I'm sure that this was not one of those. I could even make out the doors on the plane clearly. It's not just me either, several other residents have reported seeing the exact same thing. Some people got out of their houses to see what was causing the tremendous noise, too."[114]

As for reactions to those sightings, Malaysian officials quickly dismissed them. You can almost picture it—*move along, folks. Nothing to see here.*[115] And they were handily rejected with typical major media doublespeak like: "I can confirm that the Malaysian Chief of the Defense Force has contacted his counterpart in the Maldives, who has confirmed that these reports are not true."[116] Well, gee whiz, thank you very much, but you know what? We thinks, to quote Hamlet's mother, thou "doth protest too much," and the opposite is more likely to be true.[117] We trust the islanders' reports more than we trust yours.

To recap the main witness testimonies:

- Oil rig worker Mike McKay is certain that he spotted a high-flying jumbo jet that was engulfed in flames. He detailed the sighting precisely. It took place in the area where Flight 370 went off course and made a dramatic change to the southwest from its northeast flight path. If McKay is correct, this is where Flight 370 would have suffered a catastrophic event, such as taking a missile hit and/or having a massive fire, causing it to change course for an attempted emergency landing at the nearest safe landing zone.

- Multiple eyewitnesses report seeing a low-flying jumbo jet and hearing the whining of a jet engine so loudly that they reported it to police as an explosion they heard in the sky. The locations and sighting times of the witnesses align with the time and location that Flight 370 went missing and made an abrupt course change to the southwest, toward the Indian Ocean.

- Multiple eyewitness sightings in the Maldives Islands chain report a very low-flying jumbo jet. They describe the plane and it has the same colors as Malaysia Airlines' Boeing 777s, right down to the "red stripes across it."[118] The sightings take place on a plot line going west, over the Indian Ocean, exactly where the course change of Flight 370 would have put it, and it would have had enough fuel to make it there.[119]

Let's spell it out: most of the eyewitness statements line up with each other. *And the eight men, who were on the beach, heard a loud explosion right at the time that air traffic control lost contact with Flight 370.*

The eyewitness sightings have been ignored by mainstream media, which continues to parrot the official version, that no one really has any idea what happened to the plane and its passengers, but the authorities are "still looking."

Six

Media Coverage

Midwest Today magazine had an excellent article detailing how the demise of traditional investigative reporting has left a vacuum in the world of news that has led to a "fusion" of news and entertainment and a "corporate-controlled ideology substituting for facts" that is largely responsible for the "dumbing down" of the public.[120] That mainstream media vacuum has never been more obvious than in its would-be investigation of Malaysia Airlines Flight 370.

As you may have heard, a major poll by CNN found that nine percent of Americans actually believe that alien life forms are likely to have been involved in the disappearance of Flight 370, which is funny because it occurred to *us* too that Condoleezza Rice may have been involved. But seriously, and to put that in its proper context, which is often not a simple thing to do in the case of CNN, here was the actual question used in that poll:

"Now here are a few groups that some people have suggested may be responsible for the disappearance of Flight 370 if an accident or mechanical failure was not the cause. As I read each one, please tell me whether you think it is very likely, somewhat likely, not too likely, or not likely at all that the disappearance of Flight 370 was due to actions by that group."[121]

Also, interestingly, each of CNN's pollsters was also instructed parenthetically:

"(RANDOM ORDER A TO C, ALWAYS ASK D LAST)"[122]

Quite possibly, there are psychological profilers at CNN's law firm who know the true reason for this; but if we had to guess, we'd say it has something to do with the fact that "D" is the wild card answer: "Space aliens, time travelers, or beings from another dimension."[123]

In any event, 3 percent of those polled responded "very likely" to the alien question, that had to be asked last, and 6 percent answered "somewhat likely," making nine percent who think "that group" was involved. At least that's what you got when you let the folks at CNN do the poll in *just such a way*.

In fairness, 57 percent of people surveyed responded that whatever happened to the plane was "very likely" or "somewhat likely" due to "terrorists or people associated with a hostile foreign government." Forty-two percent responded that it was "very likely" or "somewhat likely" due to "hijackers not associated with a terrorist organization." And 66 percent responded that it was "very likely" or "somewhat likely" due to "the pilot, copilot, or another member of the plane's crew."[124] When you look at *those* numbers, suddenly the people taking the poll don't seem so imbecilic after all. But it was the nine percent on aliens that received all the press coverage. Great journalism, just the type of intentionally misleading information that most of us have now come to expect from mainstream media.

If you're interested, you can see the full poll results of that question at http://nymag.com/daily/intelligencer/2014/ 05/cnn-poll-did-aliens-make-flight-370-disappear.html.

Of course, no one tells us whether this poll took place at a critically acclaimed university during morning coffee break time, or in the parking lot of a liquor store near a Star Wars convention in Las Vegas at four o'clock in the morning. We're just supposed to have faith in the people who constructed and conducted the poll and never *ever* ask them why it was such a big deal to always ask *that* question last.

Fortunately for CNN, the poll didn't include the question, "Would you agree that, in your entire life, there has never been seen such a ridiculous circus masquerading as actual news coverage?" *Very* likely? *Somewhat* likely?

We're not going to list all the journalistic atrocities in CNN's coverage, but if you have the stomach for it, you can find them in an article appropriately entitled "CNN's 9 Most Deplorable Malaysia Airlines Flight 370 Moments," http://nymag.com/ daily/intelligencer/2014/04/cnns-9-most-deplorable-flight- 370-moments.html.

Nevertheless, ladies and gentlemen, and you, the reader, too— as a direct and proximate result (just to show CNN that *we* know lawyers too) of the response to that alien question, we will include that possibility as a chapter in this book and *without sarcasm*—well, not much anyway.

CNN received well-deserved criticism for its "breaking news" Malaysia 370 coverage, which lasted for over a month, regardless of whether there was any news breaking or not (there usually was not). When things got slow, they used that strangely worded poll to convey to viewers that nine percent of Americans believed it likely that aliens were involved in Flight 370's disappearance. As we explained, the polling results were not as simplistic as that, yet CNN used that slanted result to open the door, so that they had fresh new angles of "news" coverage on the "black hole hypothesis," a "zombie plane," and parallel universes. One of their anchors even "put it out there" (his words, not ours) that maybe something "supernatural" was responsible.[125]

But the poll and its effects were just the tip of the iceberg. CNN's ceaseless mindless commentary on the flight caused its network television counterparts to point out the absurdity of it. On an episode of *The Daily Show*, Jon Stewart summed it up, "Let's recap CNN's last week of the missing Malaysian airliner coverage: There is a lot of non-Malaysian-airliner garbage in the ocean; the heavier of which, sinks—the lighter of which, floats. Let us go now to Wolf Blitzer in the new Duh Room."[126]

And *The Daily Show* wasn't through yet. When Stewart showed a clip of CNN doing a piece on how one of its correspondents had sat in front of a flight simulator for a Boeing 777 for many days (poor baby), Jon Stewart yelled out dramatically:

"*There it is!* With a lack of new developments, *CNN—is about to do a story—covering their coverage—*of the Malaysian plane story."[127]

Comedian Bill Maher of *Real Time with Bill Maher* was even less kind:

> Watching CNN continue to breathe life into this thing is like watching a doctor on *Grey's Anatomy* pounding on a patient's chest until another doctor has to pull him off and

say, "Derek, it's over." That's what I want to say to Wolf Blitzer. Wolf, it's over. Time to move on. There'll be other ocean disasters. We'll always have Atlantis, but this isn't news anymore. It's an episode of *Unsolved Mysteries*. Now, I'm sorry. If you're still glued to CNN for this "breaking story," you aren't a caring person. You're not a caring person unraveling a mystery. You're just a ghoul who's sitting on the remote.[128]

But hey, *those are comedians*, you might say; they're *expected* to do that. Well, as Jackie Gleason used to say, *"HAR de HAR har."* The comedians were *kind* in comparison to those reviewing the state of the media. Check out what a few of *them* had to say:

CNN has become a laughingstock and the butt of many jokes over the last month for their breathless 24/7 coverage of Malaysia Airlines Flight 370. Although the plane disappeared on March 8th and is still missing—presumably somewhere at the bottom of the Indian Ocean—CNN devoted non-stop coverage to the story, even when no new news was being reported.

Although CNN was late-night fodder for its discussion of black holes swallowing up the plane, it is in all seriousness a troubling sign for the American news media.[129]

Here's another beauty:

"CNN's coverage of Malaysia Airlines Flight 370 has been groan-inducing for weeks now. No matter what else is happening in the world or what time of day it might be you can bet that if you turn CNN on, you're guaranteed to see something uninformative and plane-related. The network has been roundly criticized for its wall-to-wall coverage for good reason.[130]

And *more*:

CNN's coverage of Malaysia Airlines Flight 370 has reached such astounding stupidity in the past five weeks that it accomplished something previously unimaginable: uniting Bill O'Reilly and Rachel Maddow on the same side of an issue. Both cable news hosts think their counterparts at CNN should shut up about the missing plane—not that [President of CNN Worldwide] Jeff Zucker cares. CNN's ratings are through the roof, and if it takes blind specula-tion or goofy props to maintain those numbers, then—dammit!—that's what viewers will get. It's been 39 days since Flight 370 disappeared, and we still know little about what happened, but CNN keeps talking, and in doing so, embarrassing itself.[131]

When Flight 370 was reported missing, most of main-stream media simply parroted the official explanation of the missing plane and passengers to a shocked public. The con-stant "breaking news" was that something had caused Flight 370 to go "invisible" and the location and any knowledge of what actually happened to the plane was simply unknown. They speculated that one of the pilots had turned off the tran-sponder in the cockpit of the plane and, for that simple reason, everything about the plane and its whereabouts had become a great mystery.[132] We quoted this earlier, but can't help but repeat it: CNN, a leading "news" source, actually quoted one of their aviation and airline correspondents, Richard Quest, to note that with the transponder off, the plane was "flying blind from the ground's point of view."[133] A giant airplane invisible to people on the ground. Sure, that makes sense.

As we said, much of the public found the official line difficult to swallow. How would this make any sense given all our satellites and technology? If just turning off a transponder makes a jumbo jet go completely invisible to all our technologies, then aren't we all in a heck of a lot of trouble? Even if the implication that some sudden lack of oxygen change on the plane asphyxiated everyone on board and it just kept flying on autopilot out into the middle of the ocean, there are still *huge holes in that explanation, like where is the plane?*

This is a good example of the old expression: Don't let the facts get in the way of a good story. In reality, yes, one of the pilots *could* have turned off the transponder in the cockpit. But, *sorry CNN, that does not make a Boeing 777 become invisible!*

While mainstream media was busily propagating the myth that if a pilot turned off one transponder then the airplane magically became invisible, people who actually know about such things were conveying quite a different picture. An excellent article about this by Marc Weber Tobias, a security systems expert, was published in *Forbes*, and even the article's title is a telling indication of how much misleading "news" has been presented. "The Oddity of Malaysian Airlines Flight 370: Planes Want To Be Seen"[134] is a great review of security systems and we highly recommend that you give it a read. You can access it at www.forbes.com/sites/marcwebertobias/2014/04/17/the-oddity-of-malaysian-airlines-flight-370-planes-want-to-be-seen/2/.

So, there is no evidence that turning off a transponder will make a Boeing 777 undetectable—and there is no evidence that a transponder was turned off. All we know that relates to this and has any degree of certainty is that a transponder stopped functioning correctly.[135] That could be from any number of causes, including an explosion.

As far as making the plane invisible: *Sorry, doesn't work that way.* In fact, the evidence is to the contrary; there are a myriad of built-in fail-safe security systems in a Boeing 777 that specifically make it detectable, even if the transponder is turned off. And those other security systems cannot be turned off for the *precise reason* that even if the cockpit was compromised and a pilot was forced at gunpoint to turn off the transponder, there are always many ways to track a jetliner that big because systems specifically exist to ensure that.

Chapter One of this book details the multiple security systems a Boeing 777 has, and exactly how they work. In *addition* to the transponder, there are three emergency locator transmitters and two black boxes (the CVR and the FDR) and they are automatic—they *cannot* be manually turned off. They emit emergency locator signals for thirty days and they function underwater. That's not to mention ground radar, military radar (which is an entirely separate system that also tracks every plane out there), satellite communications tracking systems, and global infrared photography. So explain to us again, please, how a 243-feet long, 200-feet wide, and 60-feet high ultra-modern jumbo jet goes "invisible" simply because the pilot for some reason turned off one transponder in the cockpit. We think they're going to have to run that one by us again.

To convey the preposterousness of that notion, consider this: Even in the old film *Turbulence* (which you can access on

Netflix) from way back in 1997, there is a scene where air traffic control realizes that the primary radio communications system of the plane they're trying to reach may have been knocked out and that whoever is left in the cockpit may not be able to respond to ATC via normal radio. So ATC instructs whoever is left in the cockpit, which happens to be a flight attendant, that if they can hear what ATC is saying, then to confirm that by simply clicking the "squawk" switch, which would still function even if the radio did not.[136] The point is, the built-in fail-safe systems have been around for a long, long time. That movie was released *seventeen years prior* to the Flight 370 incident, and that emergency procedure was already so commonplace that it was referred to as "a given" in a film.

To put forth the notion across mainstream media that, in 2014, an aircraft as sophisticated as a Boeing 777-ER is invisible, alone, and incapable of communications simply because the primary transponder isn't functioning, is so incredibly misleading that one has to shake one's head in amazement that they can even utter the words.

It's an interesting scenario in *Turbulence* because a flight attendant is forced to land a damaged commercial jetliner. That has actually happened, in case you're wondering, although the flight attendant involved had a pilot's license and she assisted the captain; she did not land solo.[137] The event is referred to as a "talk down landing," but—according to Wikipedia, at least—there have not been any, in reality, that involved a large commercial jetliner:

"There is no record of a talk down landing of a large commercial aircraft. There have, however, been incidents where qualified pilots traveling as passengers or flight attendants on commercial flights have taken the copilot's seat to assist the pilot. Modern airliners have autopilot systems that should be able [to] land a plane by themselves."[138]

Of course, remember, that's from Wikipedia, not a pilot. Ask a pilot and you'll be told that a plane requires substantial input from the cockpit controls in order to land safely.[139]

Even in *Turbulence*, it was the flight computer that actually landed the plane, with those preselected inputs from the flight crew programmed into the autoland system. The pilot, or another member of the flight crew, programs the data on the control display unit (CDU) screen by selecting the proper directives from a continuing selection of choices from the flight computer.[140] For example, in the film, to land the plane at Los Angeles International Airport, the selections were "Depart-Arrive" then "KLAX" followed by "2-5-Left" for the runway selection and then "EXECUTE"; and the flight computer put that baby down on 2-5 Left. Then they lower the landing gear and ground control says, "Leave her alone. She'll do the rest all by herself."[141] Neat, huh? And that was almost twenty freaking years ago for a Boeing 747. It's a hell of a plane and it's hard not to have a lot of respect for the men and women who engineered it. And the Boeing 777 is an even *more* sophisticated piece of engineering.

Remember this important fact: Planes want to be seen. All their systems operate redundantly to make sure they remain *visible*. If reports are correct, then the fact that Flight 370 apparently went "under the radar" to whatever extent that is currently possible, is a clear indication that someone or some ones, for whatever purpose, were trying to *avoid* being seen.

If you're looking for a good scare, go to the CNN homepage and take a gander at some of its headings for articles on

Flight 370. You'll see its negative position on the incident. Here are some of the titles:

"Sometimes you never find the crash site"

"Pilot: Why Flight 370 may never be found"

"Nine aviation mysteries highlight long history of plane disappearances"

"Why are we so gripped by missing Malaysia Airlines Flight 370?"[142]

How about this for a quick question, CNN: What the hell is really going on with all this?

Even CNN, by the way, concedes the following points:

"Hijackers or renegade pilots cannot disable some of the emergency beacons, namely, the ones attached to the plane's airframe. They are powered by batteries and inaccessible to the crew. So by all accounts, the attached beacon on Flight 370 should have activated if the plane crashed." The article asked, "Why didn't Flight 370's emergency beacon work? Why didn't the beacon send a distress signal to satellites overhead?"[143]

Here's another question for you, CNN: What happened to all the known methods of tracking a huge commercial jetliner?

And then, if we may, one final question, *Where's the plane?*

Seven

Public Reaction: Maybe There *Is* Hope After All

The first thing you'll notice if you start reading up on Flight 370 in the mainstream media is that readers' responses to the news coverage are quite a bit sharper than the news coverage is. Maybe major media companies should include the college class *An Introduction To Logic* along with their journalism requirements. It's apparently needed.

A *Reuters* article, for example, was published online months after the disappearance and seems to give credence to the Australian's government claim that everyone on the plane probably died from hypoxia (oxygen deprivation).[144] There were more than *four thousand* reader comments in the first twenty-four hours the article was up, and most of them were refreshingly logical, especially when compared to—you guessed it—the article on which they were commenting. The article has since been closed for comments. Maybe they were too intelligent. There were some real gems in them too. Below are just a handful that we pulled at random.[145] We include them in all their glory; original syntax and grammatical errors are expressly *not* corrected. And they are not what you are used to hearing either. Instead, they are *raw*, *honest*, and *logical*. We recommend going to the site to read more.

John C: "the plane was deliberately diverted thousands of kilometers from its scheduled route before eventually plunging into the Indian Ocean.

So it was deliberately diverted, then set on auto pilot so they could suffocate and crash? And we know this without any evidence except the satellite images that show us where the plane went, yet we can't find the plane. Crystal clear now.

Gary: "hard to believe, I flew in the air force and was trained repeatedly every year to recognize hypoxia systems in an altitude chamber, plus the aircraft warning systems would have let them know."

Dimka: "To the Lead Investigator: And yet there is not a single piece of debris anywhere? As a lead investigator you should know that ripped peaces of that aircraft should have land into to the ocean and float miles and miles across. Three and a half month later we still have nothing. Where is it? Don't tell me that in went into to the water as whole piece. Where are parts of fuselage, a like a fin, parts or whole wings, luggage, bodies, life vests, and thousand more items that would float, which were sucked out while it was going down. In 3 month they haven't found #$%$.

Which means it was flying around to confuse people like you, then it was landed in some remote area and I hope I'm wrong 100%, but I hope someone is not making a long range missile out of it. If you are lead investigator god help us all."

TimBrock: "HARD TO BELIEVE THIS THEORY. An airliner falling into the ocean woud hit at such a speed, it would shatter or explode on impact. hitting water at that speed is like hitting a brick wall.

So, if the plane crashed into the ocean, it would have exploded into tens of thousands of pieces. Many of those pieces made from luggage (or contents of luggage), insulation, seats, aircraft tires, etc., etc., would have floated and at least a few pieces would have been found. But still, not one piece, not a fragment, nothing of or on that plane has been found!

Heres what I think is more likely: Terrorists hijacked the jet, either forced the pilots to fly to an undisclosed location or killed them, and flown it to an undisclosed location themselves, landed, forced the passengers off, killed them (burried them in a mass grave), and now have the plane hidden as they load it with explosives.

Something to make 9/11 look like a church picnic.

After all, the planes transponders were switched off! That can only be done deliberately and by the pilots!"

Crackerjack: "I'm sorry, but I'm not buying that. They continue to theorize that the plane crashed in the ocean. If that were the case a plane that size would have exploded into a million pieces upon impact. There should have been at least suitcases, floatation devices, plane fragments, or something floating everywhere for miles. Yet, search after search found not one piece of anything to indicate the plane crashed there?

I can honestly say that I have no clue what happened to that plane, or its passengers, and probably never will, but I really do Not believe that plane crashed into the ocean, but do believe that there is A lot more to this story that we will never get the answers to."

Dorrell: "This is another dumb case of conjecture. It only makes things worse for the families. If this were the case, why were different communication devises shut down at different times? Why did they not call out there was an issue before hand?

Real question is, is the plane on land, hidden? Or under the sea? Until that can be answered, wit facts, be quiet."

REPENT: "And nothing floats, yes its a big ocean yet a bottle tossed carelessly into the ocean 10 years ago can and do wash up on beachs around the world, stuff should have washed up some place."

Pete: "I'd love to know how a plane that size could crash ANYWHERE in an ocean and NO WRECKAGE was found ANYWHERE! It just doesn't land gracefully in an ocean and then sink!"

Vince: "why was the transponder turned off? If you are suffocating that would be the last thing you would do. Australia does not build airplanes so how would they know?"

McCloud: "Big plane doesn't gracefully alight on the ocean's surface and sink intact. It breaks apart violently and things come out. Show me debris. A toothpick. A napkin.

A pice of plastic. A seat cushion. Not ONE item has been recovered. That is strange."

MLKafir: "Why no debris from the aircraft? Why no bodies found floating? Something would've come to the surface by now regardless of how big the Indian Ocean is. On some beach somewhere there is debris and would be detected."

Nomsa: "They know what happened to that plane! This is such a poor cover up!!"

David: "Print facts please, this is pure speculation and should not be allowed to be printed as news. No one knows what happened to that plane and until someone does, please leave speculation to the birds! I only care about facts. Gossip and hearsay is useless."

Happily Retired!: "In short, they have no idea what happened."

Ali: "I don't buy it. I believe there is mystery behind Malaysia jet. How in the world big aircraft went missing for several month and no one has damm clue about it."

Robert: "They can't even find the darn plane, how can they speculate what happened to the passengers. Every week it's in a different location."

Daniel: "I no longer believe a word they say about this plane's disappearance. Every day their story is different."

Dirtyscrappymcfilthy: "Wow. All guesstimated with no hard evidence. Might as well say god vaporized it with his hands and sent the plane to heaven."

RonW: "Nothing about this whole boondoggle adds up at all. It just doesn't."

jimimac: "Boeing 777 aircraft come equipped with two life rafts that each carry a gps device that is attached to a sat link device so that the rafts do not get lost. If a signal was not received from one or the other of these rafts, there was no crash at sea. Boeing knows this and will not say because of pending lawsuits."

OnTheGround: "What is up with this? There is not one shred of evidence that MH370 crashed at all, or that the passengers are not still alive somewhere. The actual evidence – eyewitness accounts of a jet flying low over the Maldive Islands toward the island Diego Garcia – was dismissed by the "experts" because islanders could not identify the type of jet it was at 6 am in the morning! Do I want these "experts" on my team? – I should say not!"

It should be noted that the above comments were made after Yahoo picked up the article on its news feed. *Reuters*, which apparently does not welcome a great deal of intelligent debate about the absence of intelligent debate in its articles, simply cut off discussion of it on its site. A few days after the article came out, the folks at *Reuters* wrote, "This discussion is now closed. We welcome comments on our articles for a limited period after their publication."[146]

Sounds a lot like one of those recordings you get, doesn't it? When they tell you "Your call is very important to us," even though they just showed you that it obviously isn't!

See what we mean though? Don't those readers' comments make a lot more sense than what we're used to hearing about the plane's disappearance on television news?

For a sophisticated discussion of the issues surrounding Flight 370, read the comments on the Flight 370 page of Patrick Smith's website, Ask the Pilot, www.askthepilot.com/ malaysia-airlines-flight-370/. On it you will find comments, such as the following, and many are from experienced pilots. As with the other comments, they are reproduced exactly as they are on the site.

Jeff Latten: "Rachel, you make several good points: landing on water is like landing on green concrete, except it's wet. And, low wing-mounted engines might certainly cause the cartwheeling process you describe. Under that theory, a plane like a 727, with its fuselage and tail mounted engines has a better chance of landing in one piece. Anyone care to chime in on that?

But the more important idea is that if the plane broke up, as is likely in a forced ditching in the open ocean, and even if every part of the plane sank, there is still a huge amount of floating debris that comes off the plane; seat cushions, luggage, clothing, personal items, blankets, bodies, insulation, plastic panels, etc. The idea that all this stuff is nowhere to be found really argues against a sea ditching, but where the hell in the suspected area (or the max fuel range) would/could you land something this big without someone noticing? You can't land a 777 on a grass strip or road or field or beach, or even a little local airport designed for Cessna 172's. I'm sure Patrick knows what the minimum runway is for a loaded 777 at that temp and altitude is, but it wouldn't surprise me a bit if it was 7–8K feet or more. Also, if you're doing this at night without ATC landing guidance then runway lights would be a good thing. Just where do you have all that and no radar?"

Jo-anne: "What about all the passengers with cell phones? If I thought my plane was having trouble I'd be sending out all of my "goodbye, I Love you" messages.

Maybe no cell signal at first but for 500 miles? Just seems odd to me not one call or text rec'd from passengers. Are there On board telephones that crew could have used?"

mickey: "Indeed that is why every single one is absolutely dead, barring perhaps a lone miracle or two who quite possibly could have survived the initial crash. There is zero chance the plane could have been sucessfully hijacked or otherwise safely landed, as somebody would have gotten off a message."

Donna: "I was wondering the same thing..not ONE passenger of flight attendant tried to get a call out after they relized they weren't going to China & assuming that one of the pilots was flying, why didn't they try to talk to them? What were the passengers told, if anything? Someone would definitely have used an electronic device to send the word out something was horribly wrong."

Yvonne: "If malaysian military tracked an unknown aircraft over their country wouldn't they send fighter planes to investigate? Surely they wouldn't just watch it.......
Maybe they shot it down.

Today they acknowledged that there was a radar sighting of something but couldn't say what it was.

I smell a cover up here."[147]

So, as far as commentary on the merits of mainstream reporting, or the lack thereof, we'll give the last word not to the writers, but to a couple of readers, because that's only fair. Note that the first is actually agreeing with the comments of another reader, *not* with a mainstream article:

Charles Weinacker: "…I agree. Not in the water… Water would have surrendered evidence, long before now. Hijacked and sitting, promise."[148]

And here is the post that he was apparently in agreement with:

Kevin McConnell: It seems pretty obvious that, if the plane was flown by a human, and not autopilot, that it was stolen. The track(s) of the plane tend to indicate that it was indeed flown by a person, as it's flight path appears to have been intelligently controlled to avoid radar detection. The transponder was manually shut down. The comm radios were manually shut down. Early reports showed it climbing to a very high altitude, which I suspect would be to incapacitate the passengers by turning off cabin pressurization (the cockpit has a separate system), and then dropping below radar. It makes absolutely no sense that anyone would go to all that trouble and then just fly out to sea to exhaust the fuel and crash. I would bet the farm (if I had one) that the 777 is in a hangar getting repainted. There ARE airstrips in the world that a 777 could land on intact and not be detected. As for the CVR and the FDR, the plane could have landed briefly, taken on more fuel, removed the "Black Boxes", then continued on a path over the areas being searched, and somehow jettisoned the CVR & the FDR, and then continuing on it's way to wherever it is now. There are a couple of reasons why no-one has claimed responsibility. First, it may have been stolen to be used as a drug running plane. Second, it could be filled with explosives (or worse) and used by terrorists to do something VERY evil.[149]

Eight

Anomalies: What's Wrong with This Picture?

A lot of things seemed (and continue to be) different—to word it lightly—about Malaysia Airlines Flight 370. Its story is unlike previous air disasters, and as the facts came out, people were surprised to note a number of disturbing anomalies:

NO DEBRIS FOUND

When Malaysia Airlines Flight 17 was shot down over the Ukraine on July 17, 2014, it was a case of a Boeing 777—the same model as Flight 370—being struck by a missile and then striking hard ground at high speed. One might think it would have completely disintegrated on impact. It did not. There were still gigantic pieces of debris, simply because the triple-seven is such a huge airplane.

Flight 370 reportedly went down over the Indian Ocean. If you think all the pieces of the plane would sink to the bottom of the ocean if it did, then you better think again. You might remember that after the tsunami in Japan, pieces of debris as large as refrigerators and motorcycles washed ashore at beaches thousands of miles away.[150]

It is considered highly unlikely that Flight 370 landed smoothly on the ocean and then sank. You may think it's easy to land a jetliner on the ocean. The emergency landing that became known as the "Miracle on the Hudson" was not at all a typical water landing. As one pilot bluntly put it, "There's no such thing as a water landing. It's called crashing into the ocean."[151]

Debris following a water impact is normally dispersed in a very wide field. There are three million parts in a Boeing 777-200 ER and at least a few hundred thousand of those three million parts should have broken apart.

Here's an example: When Swissair Flight 111 crashed into the ocean on September 2, 1998, the aircraft broke up upon impact with the water.[152] Most of the debris sank to the ocean floor, but there was still a large debris field that stayed afloat in the crash area, and a lot of debris washed up at beaches over the following weeks.[153] That plane was a McDonnell-Douglas MD-11, a wide-body jet airliner that can hold 277 passengers and is similar in size to the Boeing triple-seven.[154] So if—as we have been told—Flight 370 went down over the Indian Ocean, huge pieces of it should have washed up on beaches by now, and other pieces of it should be floating. Yet we're told that nothing has been found.

NO "MAYDAY" DISTRESS SIGNAL

Typically, one of the first things that a member of a flight crew of a commercial jetliner would do in an emergency situation of any kind, is immediately radio air traffic control that they have a problem. If a situation such as fire or decompression occurs in the aircraft, one of the flight crew members communicates that very quickly, and with good reason: they require the assistance of air traffic control in order to change altitude levels safely, obtain priority over other flights due to the emergency, and prioritize them for an unscheduled landing. So it behooves them to inform ATC of their emergency at the earliest possible moment—and they are trained to do that, almost automatically.

In *every other similar case* we looked at—as you will see in Chapter Nine—an emergency was quickly radioed to air traffic control because radioing an emergency is *what they are trained to do*. In the case of a fire or smoke in the cockpit, their oxygen masks filter out smoke, and the special microphone they contain automatically activates. All they have to say is "mayday, mayday, mayday" and help will be on the way. But in the case of the Flight 370 crew, they didn't, which defies logical explanation. It's true, A-N-C—aviate, navigate, communicate—but it's also true that it's imperative to let air traffic control know you have an emergency, and at the earliest possible moment.

Typically, an emergency is quickly announced by a simple radio call from a flight crew to air traffic control. In the case of an extreme event, the call is known as "calling in a mayday" because those are the designated words for declaring an emergency. Any member of the flight crew would radio "mayday, mayday, mayday," and then give the troubled plane's name, for example, "Malaysia Three Seven Zero." This would be followed by a description of their emergency and what they are requesting, such as immediate approval for an altitude change and priority status for an emergency landing. In other in-flight emergencies, pilots have sometimes simply stated that they are "declaring an emergency."[155] Either method is acceptable, though stating "mayday" three times is considered the correct method. So they also could have said "This is Malaysia Three Seven Zero, we are declaring an emergency. Repeat, we are declaring an emergency." Air traffic control would respond immediately, offering whatever assistance was required. If it was a matter of a lesser emergency, for example a slight amount of smoke that they presumed could be easily dealt with, they would still immediately report that fact. After the problem is discussed, ATC would ask the captain if he or she wishes to declare an emergency.

It has been observed that, for whatever reasons, there is sometimes reluctance on the part of pilots to formally declare an emergency.[156] There is, therefore, another method in aviation communication to advise of a problem that the crew is attempting to correct. That situation of a "developing emergency" is communicated to air traffic control via a radio report of "pan-pan" instead of "mayday." For example, one would say "pan-pan, pan-pan, pan-pan, this is Malaysia Three Seven Zero, reporting a problem." Note that—and this is direct from a commercial pilot—*it only takes about two seconds* to call in a mayday or pan-pan distress call.[157] So, while it's true, "aviate, navigate, communicate" is the order of trained reaction, it is not realistic to posit that there was no time to communicate a problem on board the aircraft. As one "long-haul" commercial pilot puts it, "If an emergency is severe, any of the two pilots can make a very quick mayday call at any time such as 'Mayday, mayday, mayday, MH370, smoke on board, stand by.' It would only take two or three seconds to transmit … The crew may have been dealing with an electrical fire and began turning off the electrical systems which supply power to the transponder and ACARS system, but as previously stated it would take them two seconds to say: 'Mayday, mayday, mayday, MH370, electrical fire, stand by.'"[158]

TRANSPONDER "TURNED OFF"

Although we were told repeatedly in mainstream media that the transponder of the plane was "turned off," we do not know that to be a fact. We only know that, at a certain point, the transponder ceased to function. That could have been due to an in-flight emergency, such as an electrical fire, in which crews are trained to turn the transponder off. There are also other reasons they may have turned it off, only *one* of which is being under threat from hijackers. There are also, of course, things which will stop a transponder from functioning properly. A ground-to-air missile will do it. So will a bomb or a large fire.

The important thing to remember is not to make false assumptions.

For example, it is widely reported that the ACARS data transmitter (colloquially referred to as the transponder) was "shut off" when all that is really known is that it "went off."[159]

But again, communication is the key, and the absence of communication from Flight 370 must be kept in mind. As an experienced commercial pilot who has needed to turn off a transponder during a scheduled flight observed, "I can't imagine a situation where you would turn the transponder off and not tell air traffic control that that's what you were doing."[160]

VIOLATIONS OF STANDARD EMERGENCY PROTOCOL

Obviously, there are standard procedures in place to immediately enact the moment a commercial flight goes off radar, and *doing nothing* is not one of them. When 370 went "dark" — not appearing as expected by Vietnam air traffic control after the flight was handed off to them by Kuala Lumpur air traffic control—Vietnam ATC apparently acted properly. The closest flight to where 370 had been—presumably—was contacted and requested to radio Flight 370. In fact, we learned—via Simon Gunson—that at least two other commercial flight pilots in the vicinity of Flight 370 tried to contact it.

According to Gunson:

> The crews of JAL750 and MH88 (a Japan Airlines flight and a Malaysia Airlines one, both of which were the closest two planes to Flight 370 that air traffic control could contact) did have garbled voice communication with MH370 on 121.5 MHz, the international distress frequency, after 1:30 A.M. MYT. The captain of JAL750 said he spoke with the copilot of MH370 just after 17:30 UTC (1:30 A.M. MYT) when JAL750 was climbing northeast from Ho Chi Minh City (Tân So'n Nhât Airport).
>
> The Malaysian newspaper *New Straits Times* published the interview with that captain, however, it has recently removed the article after I pointed out that it was impossible for MH370 to speak with JAL750 at that time if the Malaysian airliner was flying west from IGARI.

Unfortunately, *New Straits Times* has published a load of trash trying to blame Captain Zaharie Ahmad Shah and does not like people ridiculing their theories. That is why the article has been removed, because it disproved pilot suicide.[161]

Those efforts by air traffic control to establish contact via another flight failed. So Vietnam ATC then got back in touch with Malaysian authority, not once, but *twice*; informing them that Flight 370 had *not* entered their airspace as had been expected. At that point, security protocol would have been to declare an all-out emergency, even sending a fighter aircraft to locate the plane and escort it to safety. That was reportedly *not* done, although there have been reports that the plane *was* observed with a fighter escort.[162]

The immediate scrambling of fighter jets should have occurred *because scrambling fighter jets is the security protocol* in that precise situation. It's in everybody's interest because the pilots in the jets that intercept the plane can see what's actually going on and, hopefully, escort the airliner to a safe landing. That's how it's done. As but one example, in 2005, when Helios Airways Flight 522 became non-responsive above the mountains of Greece, two F-16 fighter jets were scrambled within minutes.[163] They intercepted Flight 522 and established visual contact *so close* to the cockpit that they could actually see the first officer slumped motionless over the controls. They determined that the captain's seat was empty and that the plane was flying on automatic pilot, and even saw the oxygen masks dangling in the passenger cabin.[164]

The point is this: the notion that sometimes "there's nothing you can do," which has been the general impression conveyed to us regarding Flight 370, is not simply wrong, it's *way beyond wrong*. There are **always** things you can do and, moreover, there is an established process of security protocols to follow that you ***do follow*** in *every* emergency situation—unless for some reason you absolutely cannot.

It becomes clearer as we learn and assess the circumstances in their entirety that we have not been accurately informed by authorities of what actions were actually instituted at the phase of what *should have been* a massive crisis situation with a huge commercial airliner. They *should* have sent up fighters; by some reports they *did* send up fighters,[165] but, for whatever reason, they're not telling us that they sent up fighters.

If Malaysia's *official* reports are to be believed—and, frankly, it would be quite a stretch to do so—a full seventeen minutes went by before they realized that 370 had disappeared from radar.[166] But, like we said, that's "if reports are to be believed." This whole thing gets to a point where you really have to begin to question the truthfulness of some of these reports. As one reader commented in response to a CBS article, "The story doesn't 'wash' because of the heavy military presence in the region. No way. US Navy alone had two Arleigh Burke class destroyers that should be able to track this plane easily. Australia flies surveillance/spy planes from Australia to Malaysia. US military has radar arrays on land in Indonesia and Singapore."[167] And yes, there again, is another case of a reader comment that is more intelligent and logical than the supposed "facts" from the experts. At what point did they leave logic behind in this case? From the very beginning?

According to the CBS article, "The report also said Malaysian authorities did not launch an official search and rescue operation until four hours later, at 5:30 A.M., after efforts to locate the plane failed."[168]

Unlike mainstream media, the families of the missing passengers have some very intelligent questions. They formed a committee, the Committee of MH370 Family, and they have presented authorities with a list of intelligent questions.

As a public service to the family members of those individuals on Flight 370, here are the questions they have demanded answers to, and, as of the publication date of this book, to which they have not received adequate answers. They were posted on the Chinese site Sina and are reproduced here exactly as they were written:

KEY QUESTIONS FROM FAMILY MEMBERS

Emergency locator transmitter

As far as we know, MH370 has one fixed ELT and two portable ELT. All ELT have passed the latest maintenance check (Malaysia airlines has promised to ask when they did the last check and what has been checked)

1. How many ELT are there on the plane? Including fixed and portable. We have heard two versions. We would like to insure how many?

2. Did Malaysia Airline have regular maintenance checks for ELTs? When was latest check for MH370's ELT? What is the maintenance interval? And we need to see results of maintenance check. If not possible to see the result during the investigation. We would like to know what has been checked.

3. How your 406 MHz is certified. Your licenses?

4. Is it possible to break the ELT at high impact? Where is the 406 MHz ELT exactly located on MH370? (tail of the flight or on the ceiling of business class)

5. Is the ELT protected in the compartment within the fuselage? Surrounded by metal? Will the signal be weakened by the metal shield?

6. Is the cable and blade antenna 9G certified? How much impact is needed to activate ELT? Did the ELT activate in Asiana accident and France Air 447?

7. Are 121.5 MHz 243 MHz useless when aircraft crashed in water?

8. The manufacturer of the ELT, the signal of 406MHz supposed to be detected by satellite.

9. Is the 406 MHz a separate service beside the 121.5 and 243 MHz?

10. If the crew has been trained on how to use ELT?

11. Can ELT unlock and bounce to the surface of water?

12. When plane is trying to land on the sea, can ELT be activated?

Black box

1. The serial numbers of the black box on MH370. Manufacturer?

2. What kind of characteristic signal does MH370's black box pinger send? Is it a pulse, with a peak at 37.5 kHz? The width, the shape of the pulse?

3. What is the sample that Boeing has sent to Australia to compare with the detected pinger? Was it from a normal black box signal or the specific black box on Boeing 777, or even the exact black box mount on MH370?

4. How many items can the Flight Data Recorder provide? 25, 57, or 88?

5. How long time can the Cockpit Voice Recorder provide? When did Malaysia Airline start to prolong 2 hours recording?

6. Can the detected frequency 33.3kHz illustrate the coverage environment of the black box? Can the location of the black box also be illustrated?

7. Can the investigation team make an experiment on checking the 33.3kHz is caused by the weakened battery? And how long can 33.3 kHz be detected under the weakened power situation?

Protocol

1. What protocol does ICAO do when flight missing? What did Malaysia Airline do when MH370 missing? What organizations are Malaysia in?

2. We want MH370's logbook

3. We need Malaysia Civil Aviation Control MH370 Voice Record.

4. Inmarsat in Malaysia, NTSB Chief engineer, Zaharie personal contact phone number. Direct contact information.

5. Have the searching and rescue team got final result from the searched areas? Are they sure any impossibility on those closed area, if not, why close all the other areas?

6. Can Malaysia Government specify the right of kins, especially the right to know the facts of a case or the details of an incident?

7. We require the ATC (Air Traffic Control) audio.

From:
The Committee of MH370 family[169]

DANGEROUS CARGO

The cargo aboard the aircraft, over and above the passenger luggage, had some interesting items. There were over three tons of mangosteens, which might sound like a mineral, but are actually a type of exotic fruit. At first, Malaysia Airlines officially reported that there was no hazardous cargo aboard the aircraft, just the mangosteens.[170] That wasn't true, or—hey, let's put it another way—that was a lie. There was also a large shipment of lithium-ion batteries, which are highly flammable and, in fact, caused a dangerous incident involving fire on another air flight in 2010. The shipment on Flight 370 was reportedly 221 kilograms, more than large enough to be a potential danger.[171] Those dangers and that incident are covered in Chapter One of this book, in the section titled "Automatic Fire and Smoke Suppression Systems."

Also of note in the cargo was a large shipment "declared as radio accessories and chargers" that, according to the air waybill, weighed 2,232 kilograms.[172] A source at Malaysia Airlines reported that he was not permitted to provide additional information on that. We have no idea why, and they are apparently not going to tell us. His exact words were the following: "I cannot reveal more because of the ongoing investigations. We have been told by our legal advisers not to talk about it."[173] All the cargo, and even the in-flight meals, were being investigated by police, in order to rule out sabotage.[174]

ALTITUDE CLIMB TO 45,000 FEET

Another error that was widely disseminated to the public was that Flight 370, after losing communication with air traffic control, went into a steep climb that brought the aircraft all the way up to an altitude of 45,000 feet.[175] This report immediately gave birth to other widely disseminated notions, such as the one stating that the reason for doing that would be to quickly immobilize every passenger on the plane by depressurizing the plane (some went on to claim that the flight crew has special protection from depressurization) as a deliberate act by the pilot. All this was apparently, as they said, so that the insane pilot could crash the plane into the ocean just to show everybody that he was in control. Crazy, right? First of all, there is absolutely no evidence indicating that a steep altitude climb was even attempted. And second of all, our research indicates that <u>a climb like that in a triple-seven is not even *possible*</u>.

So that's quite a theory, an unnecessarily-elaborate kamikaze death climb, but, as they say in the World Series of Poker, read 'em and weep. Malaysia Airlines Flight 370 did not dramatically climb to 45,000 ft. and then dive below 23,000 ft. after completing a U-turn before it disappeared. That is the conclusion of investigators looking into the disappearance. They've even gone on to say that Malaysian radars hadn't been calibrated precisely enough to draw any conclusions about the plane's altitude.[176]

Interestingly, you can still find lots of misinformation on that topic. In fact, most people out there probably still believe that the flight made that steep altitude climb.

Simon Gunson told us that he knew immediately that the climb to 45,000 feet reported in the media was false information. Gunson explained how he *knows* that a Boeing 777 didn't make that climb; it's because the plane's computer system knows that the aircraft's "ceiling"—its altitude limit—is 43,100 ft. Gunson said:

> In fact, the flight management computer would not permit a climb above 43,100 feet and would have to be disconnected to go over that ceiling. Any climb over the service ceiling would require a degree of precision flying by hand that just is not possible. If you disconnect the autopilot, then the margin between stall speed and MACH Buffet is just six knots and no pilot is precise enough to hand-fly an aircraft within such a narrow margin. The problem is one of energy and the engines can no longer push the aircraft fast enough to overcome the stall speed. Between the last confirmed altitude 35,000 ft. and 38,000 ft., it would have taken MH370 something like fifteen minutes to climb the extra height. It is not like a fighter plane where you kick in afterburner and point the nose up. A big heavily-laden airliner is sluggish and takes ages to climb. A climb from 38,100 ft. to 45,000 ft. if achievable at all would take at least twenty minutes."[177]

And here's another pilot who confirms Gunson's logic, as he responds to the following questions:

"Could a pilot force decompression to occur and prepare for it?"

"No. Impossible."

"How dangerous is it for a commercial aircraft to fly at 45,000 ft., and does the theory that MH370 was

flown to that height in order to deprive the cabin of oxygen hold up?"

"No, it doesn't. The cabin and flight deck atmosphere at high cruise altitude generally has the atmospheric pressure of 8,000 ft. above sea level. Rapid decompression is a theory, and if the plane experienced a decompression, the crew would have to don their masks quickly to avoid passing out. But on the emergency procedure, step one in case of decompression is to don masks. We have an oxygen supply, separate systems for the passengers and crew.

"The altitude at which an aircraft can fly depends on its weight, essentially. When an aircraft takes off, fully fuelled, with a full load of passengers, it will initially climb to an optimum cruising altitude. Over time as fuel is burnt off, the optimum cruising altitude increases, so pilots perform a step climb throughout the flight as the plane becomes lighter and lighter.

There are also limitations on the maximum altitude a plane can fly for a given weight, and if exceeded there is a risk of high speed stall. If you Google 'coffin corner,' it may make more sense. In terms of the conditions within the cabin and flight deck, there would be no issue with 45,000 ft."[178]

RADAR TRACKING LOST (REPORTEDLY)

Just about everyone in the world, except for the authorities and mainstream media, has a very basic question about Flight 370: How do you suddenly lose *all* forms of communication and tracking with something as large as a Boeing 777 jetliner when something as small as a cell phone can be easily tracked? The answer appears to be: You don't. Although we have been told, and by no less than the prime minister of a highly developed nation, that Flight 370 went "invisible" after the transponder was "turned off" by the "deliberate action from somebody on board the aircraft,"[179] here are four words to remember: DO NOT BELIEVE IT. Common sense should tell you otherwise, and that's an instinct you should trust.

Take a good long gander at the subheading in this section entitled "Admissions From Public Officials," and you will see that some very high-placed officials, who were very close to this whole thing, consider the whole "lost plane" scenario about as believable as the one about the moon being made of green cheese and other fairy tales you may have heard.

CONTRADICTIONS IN EVIDENCE

Large chunks of the evidence seem to cancel each other out. In one respect, the evidence points to a sudden catastrophic event that did not give the flight crew enough time to declare an emergency, and caused the immediate failure of all communications and emergency devices on board the aircraft. Yet, in another respect, if the information reported is correct, evidence *also* reveals that the plane continued flying and *could not* have suffered a catastrophic event.

Inmarsat, the company that provided satellite tracking info, made a formal statement that, "Routine, automated signals were registered on the Inmarsat network from Malaysia Airlines flight MH370 during its flight from Kuala Lumpur."[180] It also said that the missing aircraft was equipped with an Inmarsat signaling system that should help pinpoint the plane's location via their system of satellite "pings."[181] Those statements led others to the conclusion that the plane could not have suffered a catastrophic event, because those signals would have stopped simultaneously. The company said, "If the plane had disintegrated during flight or had suffered some other catastrophic failure, all signals—the pings to the satellite, the data messages and the transponder—would be expected to stop at the same time."[182]

This begs the question, if there was no catastrophic event that knocked out communications, then why were there no communications?

FLIGHT PATH CHANGE (REPORTEDLY)

Officially, we are informed that Flight 370 departed from its northeast route to China and made a sharp turn west, taking it over the Straits of Malacca.[183] We asked Gunson about this and he doesn't think that westerly turn actually occurred and explains why in a very logical argument. His note to us is below. We've included the URLs he sent, so you can get visuals of what he's referring to:

> It was confirmed by Angus Houston—the retired head of the Australian military who is in charge of the search for Flight 370—on 24th June 2014, that MH370 was not seen on radar climbing to 45,000 feet before turning west and diving below 23,000 feet. Therefore it never turned back from IGARI at all. It turned south from the coast of Vietnam. Malaysia's government claimed MH370 dropped to 5,000 feet and flew west like a fighter plane low over Kota Bharu to avoid detection on radar. Malaysian police insist there were multiple sightings of a loud low flying jet over Kota Bharu by residents. However, two military

CCTV footage stills released by officials of the Iranians traveling with stolen passports. Fakery was obviously involved in these photos; in fact, Malaysian police even *admitted*, after they were called out, that the photos aren't as they should be. Notice how the men have identical pants and shoes, and their legs are in the exact same position. Even the shadows on the ground are *exactly the same*. The clear image manipulation rightly led some to suggest that a "terror scenario" was being created. (AFP PHOTO/MALAYSIAN POLICE)

This is the actual Boeing 777, number 9M-MRO, of Flight 370. It is pictured here, on an earlier flight, from Charles de Gaulle Airport in Paris. A Boeing 777 has three million parts, many of which float, yet no debris from Flight 370 has been found. If the plane crashed in the ocean, it wouldn't neatly sink to the bottom; pieces would wash up on shore. (Laurent ERRERA)

By comparison, when a Boeing 777 was shot down over Ukraine by a missile, a massive field of debris was spread over six miles and three Ukrainian villages. If an aircraft that large crashes into the ocean, it tends to break apart, leaving a huge debris field, much of which floats. (Ministerie van Defensie)

Swissair 111 broke apart on impact with the Atlantic Ocean. Most of the debris sank to the bottom of the ocean, but a lot of debris was found floating at the crash site and other debris was found on beaches, where it washed ashore in the following days. (Trevor MacInnis)

This is the cockpit of Flight 370's Boeing 777, 9M-MRO. Placing a "mayday" distress call takes [?]–3 seconds. "Squawking IDENT," pressing a button that lights up the plane's location on the [b]oard that air traffic control follows it on, takes even less time. Pilots and aviation experts can't [u]nderstand why the Flight 370 crew wouldn't have contacted air traffic control, unless there was [a] catastrophic event. (Chris Finney)

It is well-documented that most large aircrafts can now be controlled completely by remote, from the ground. NASA and the U.S. military have proudly demonstrated the technology required to do so. This is especially true for the Boeing 777 and its "fly-by-wire" technology. In 2003, Boeing patented the "Boeing Honeywell Uninterruptible Autopilot" system under the name "System and method for automatically controlling a path of travel of a vehicle." It stated that "The control commands may be received from a remote location." (NASA)

The amount of cell phone towers has rapidly increased in recent years, and the number of cell phone pings from a large jetliner "would be a chorus" to the hi-tech listening stations the U.S. has around the world. Multiple technology experts have expressed disbelief that the plane could not be located by those digitized signals. (aripeskoe2)

Two U.S. Air Force B-52s on the runway at Diego Garcia military base in the Indian Ocean. The runways at Diego Garcia can accommodate a Boeing 777 and there were reported sightings of jet matching Flight 370 heading in that direction. There was a deleted file for practicing landing at Diego Garcia on the home flight simulator of the captain of Flight 370. A cell phone message from one of the passengers was reportedly traced to GPS coordinates near Diego Garcia, via the software embedded in the message. (Staff Sgt. Mary L. Smith, U.S. Air Force)

ROOM 302 LOGISTIC BLDG. NO.10 TIANZHU ROAD, TIANZHU IND ESTATE	CARRIER'S LIMITATION OF LIABILITY. Shipper may increase such limitation of liability by declaring a higher value for carriage and paying a supplemental charge if required.	
Issuing Carrier's Agent Name and City	Accounting Information	
NNR GLOBAL LOGISTICS (M) SDN BHD NO. 15, JALAN BATU MAUNG DIS3PLEX FREE COMMERCIAL ZONE	<<< BDS 2.73 1516/FT >>> FREIGHT : FREIGHT PREPAID AIRPORT TO AIRPORT	
Agent's IATA Code Account No.		
20-3 1232		

PLEASE NOTIFY CNEE IMMDY UPON ARRIVAL OF GOODS. ONE CONSOL POUCH ATTACHED.
SHIPPER DECLARATION FOR DGD NOT REQUIRED
'LITHIUM ION BATTERIES IN COMPLIANCE WITH SECT II OF P.I. 965'
EMERGENCY CONTACT ... 1-800-424-9300

No. of Pieces RCP	Gross Weight		Rate Class Commodity Item No.	Chargeable Weight	Rate Charge	Total	Nature and Quantity of Goods (Incl. Dimensions or Volume)
133	1990.0K	Q	GCR	1990.0	11.58	23044.20	CONSOLIDATED AS PER
67	463.0K	Q	GCR	463.0	11.58	5361.54	ATTACHED CARGO MANIFEST FREIGHT PREPAID

The waybill listing the shipment of highly flammable lithium-ion batteries as part of the cargo on board Flight 370. (Wikimedia)

MAS 370 (Kuala Lumpur to Beijing)

PILOT-ATC RADIOTELEPHONY TRANSCRIPT

Departure from KLIA: 8 March 2014

		ATC DELIVERY
12:25:53	MAS 370	Delivery MAS 370 Good Morning
12:26:02	ATC	MAS 370 Standby and Malaysia Six is cleared to Frankfurt via AGOSA Alpha Departure six thousand feet squawk two one zero six
12:26:19	ATC	... MAS 370 request level
12:26:21	MAS 370	MAS 370 we are ready requesting flight level three five zero to Beijing
12:26:39	ATC	MAS 370 is cleared to Beijing via PIBOS A Departure Six Thousand Feet squawk two one five seven
12:26:45	MAS 370	Beijing PIBOS A Six Thousand Squawk two one five seven, MAS 370 Thank You
12:26:53	ATC	MAS 370 Welcome over to ground
12:26:55	MAS 370	Good Day

		LUMPUR GROUND
12:27:27	MAS 370	Ground MAS370 Good morning Charlie One Requesting push and start
12:27:34	ATC	MAS370 Lumpur Ground Morning Push back and start approved Runway 32 Right Exit via Sierra 4.
12:27:40	MAS 370	Push back and start approved 32 Right Exit via Sierra 4 POB 239 Mike Romeo Oscar
12:27:45	ATC	Copied
12:32:13	MAS 370	MAS377 request taxi.
12:32:26	ATC	MAS37..... (garbled) ... standard route. Hold short Bravo
12:32:30	MAS 370	Ground, MAS370. You are unreadable. Say again.
12:32:38	ATC	MAS370 taxi to holding point Alfa 11 Runway 32 Right via standard route. Hold short of Bravo.
12:32:42	MAS 370	Alfa 11 Standard route Hold short Bravo MAS370.
12:35:53	ATC	MAS 370 Tower
12:36:19	ATC	(garbled) ... Tower ... (garbled)
	MAS 370	1188 MAS370 Thank you

		LUMPUR TOWER
12:36:30	MAS 370	Tower MAS370 Morning
12:36:38	ATC	MAS370 good morning. Lumpur Tower. Holding point.. [garbled]..10 32 Right
12:36:50	MAS 370	Alfa 10 MAS370
12:38:43	ATC	370 line up 32 Right Alfa 10.
	MAS 370	Line up 32 Right Alfa 10 MAS370.
12:40:38	ATC	370 32 Right Cleared for take-off. Good night.
	MAS 370	32 Right Cleared for take-off MAS370. Thank you Bye.

		LUMPUR APPROACH
12:42:05	MAS 370	Departure Malaysian Three Seven Zero
12:42:10	ATC	Malaysian Three Seven Zero selamat pagi identified. Climb flight level one eight zero cancel SID turn right direct to IGARI
12:42:48	MAS 370	Okay level one eight zero direct IGARI Malaysian one err Three Seven Zero
12:42:52	ATC	Malaysian Three Seven Zero contact Lumpur Radar One Three Two Six good night
	MAS 370	Night One Three Two Six Malaysian Three Seven Zero

		LUMPUR RADAR (AREA)
12:46:51	MAS 370	Lumpur Control Malaysian Three Seven Zero
12:46:51	ATC	Malaysian Three Seven Zero Lumpur radar Good Morning climb flight level two five zero
12:46:54	MAS370	Morning level two five zero Malaysian Three Seven Zero
12:50:06	ATC	Malaysian Three Seven Zero climb flight level three five zero
12:50:09	MAS370	Flight level three five zero Malaysian Three Seven Zero
01:01:14	MAS370	Malaysian Three Seven Zero maintaining level three five zero
01:01:19	ATC	Malaysian Three Seven Zero
01:07:55	MAS370	Malaysian...Three Seven Zero maintaining level three five zero
01:08:00	ATC	Malaysian Three Seven Zero
01:19:24	ATC	Malaysian Three Seven Zero contact Ho Chi Minh 120 decimal 9 Good Night
01:19:29	MAS370	Good Night Malaysian Three Seven Zero

 # end of file/BIT 30 March #

The final words from the captain of Flight 370 were calm and clear. There was no panic in his communication—all seemed normal. Seconds later, the plane supposedly disappeared from radar tracking, with no indication of distress.

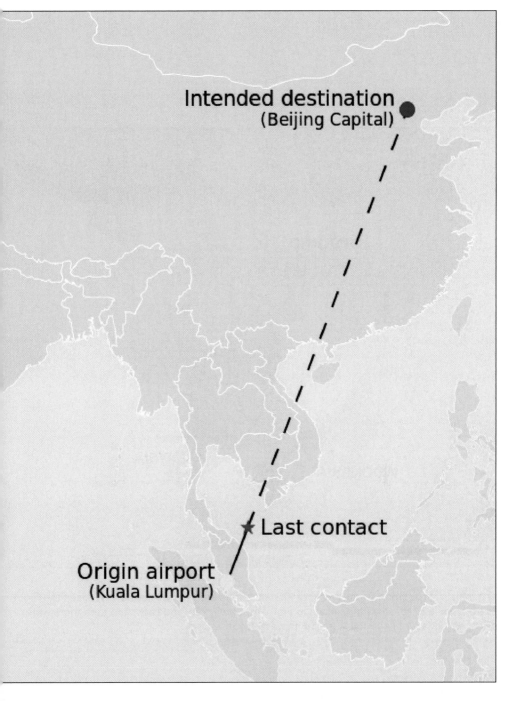

The flight plan for Flight 370 should have been from Kuala Lumpur to Beijing. (Sailsbystars, JohnHarvey)

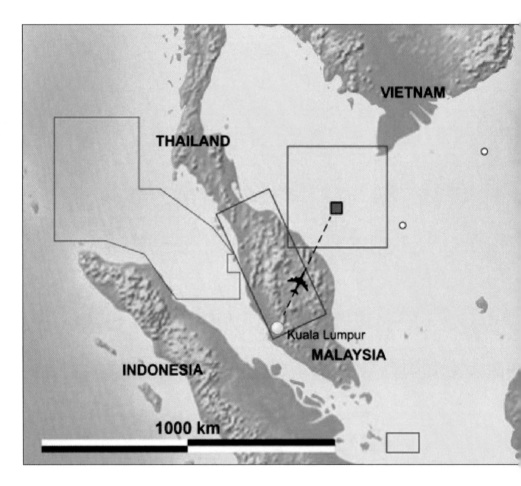

The initial search area was northeast of Kuala Lumpur, along what should have been the flight route; then land searches took place in Vietnam, Cambodia, and Laos; and then there were searches in Malaysia and the water to its west. (Soerfm)

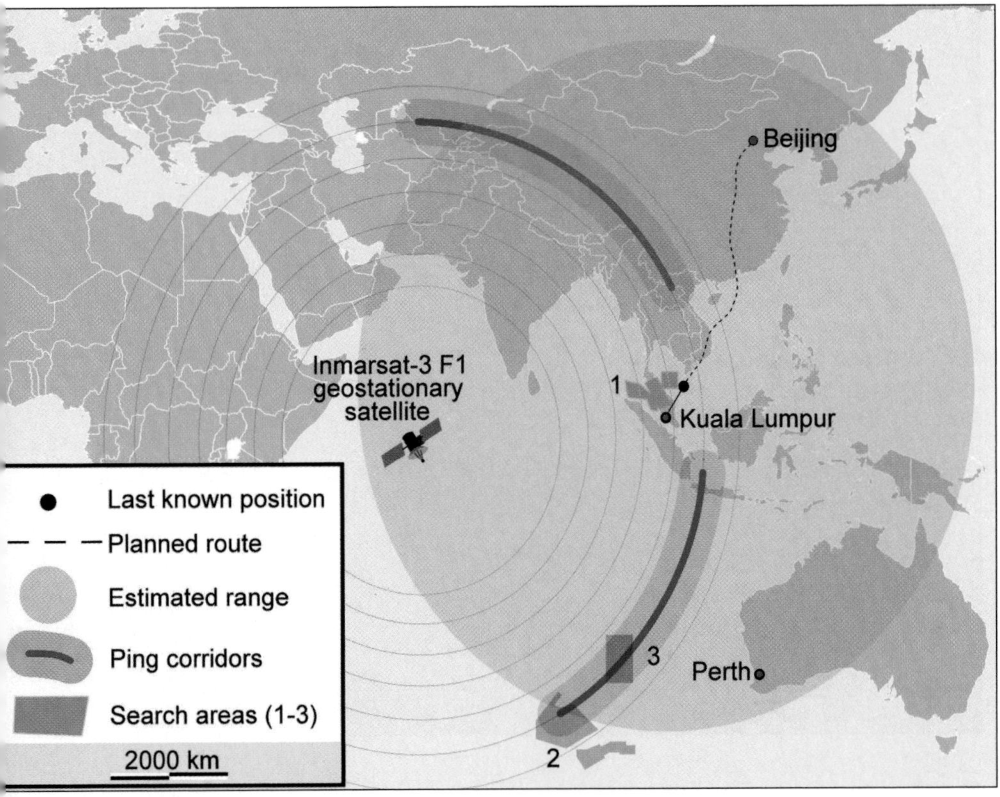
Beijing
Inmarsat-3 F1
geostationary
satellite
1
Kuala Lumpur
Last known position
Planned route
Estimated range
Ping corridors
Search areas (1-3)
2000 km
3
Perth
2

he ever-expanding search area turned into a global game of Marco Polo—and an unsuccessful ne at that. (Ain92)

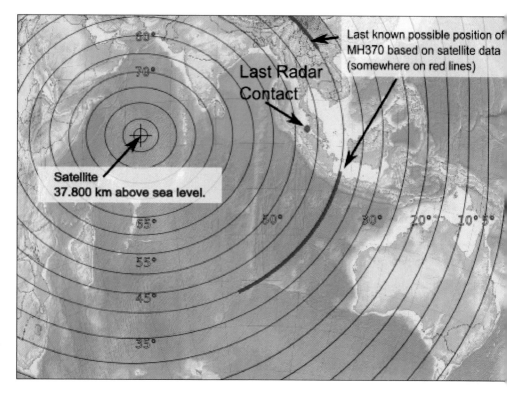

The "last radar contact" is a disputed point; it may or may not have occurred where authorities claim it did. Some information indicates that the plane never actually made the turn west. (RicHard-59)

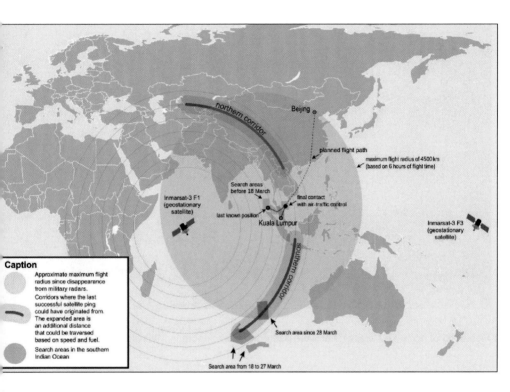

Caption

Approximate maximum flight radius since disappearance from military radars.

Corridors where the last successful satellite ping could have originated from. The expanded area is an additional distance that could be traversed based on speed and fuel.

Search areas in the southern Indian Ocean

northern corridor

Beijing

planned flight path

maximum flight radius of 4500 km (based on 6 hours of flight time)

Search areas before 18 March

Inmarsat-3 F1 (geostationary satellite)

final contact with air-traffic control

last known position

Kuala Lumpur

Inmarsat-3 F3 (geostationary satellite)

southern corridor

Search area since 28 March

Search area from 18 to 27 March

ome evidence contradicts a catastrophic event. Satellite tracking reportedly confirms that the ane stayed airborne for more than seven hours. This is another depiction of the immense area question. (Soerfm)

Messages of hope and prayer that were posted in a bookstore in Malaysia. (نف ا می)

radar stations—one at Kuantan [in Pahang, Malaysia] and another at Kota Bharu itself—should have seen this, but neither station did. This following image is from the military Thales Raytheon GM400 radar at Kuantan showing that it did have coverage out as far as IGARI, http://i257.photobucket.com/albums/hh212/727Kiwi/MH370/ActualButterworthradarimage_zps9ca910e6.jpg.

The Malaysian government also claims that Thai Military radar at Surat Thani [in southwest Thailand] saw MH370 flying west over Malaysia. The problem with this is that the radar horizon from Surat Thani would prevent the Thai radar from seeing any aircraft over Kota Bharu below 37,000 feet, http://i257.photobucket.com/albums/hh212/727Kiwi/MH370/SuratThaniradar92nmand140nm_zps1e2fcce2.jpg. Therefore the Thai radar claims are a hoax.

Next, Malaysian government officials told Chinese relatives at the Lido Hotel in Beijing on 21 March 2014 that this image was from the military radar located at Butterworth air base showing an aircraft crossing over a small island Pulau Perak at 18:02 UTC, http://i257.photobucket.com/albums/hh212/727Kiwi/MH370/Lido_Hotel_Beijing_21March_zps2eb1b3f1.jpg.

The problem with this is that MH370 could not have covered the distance at altitudes below 35,000 feet because as an airliner drops to low altitude, the increased air density prevents it flying fast enough, http://i257.photobucket.com/albums/hh212/727Kiwi/MH370/PalauPerak_zps6bb9d4eb.jpg.

Next, the image shown to relatives was not taken from a real Thales Raytheon GM40 radar, but is entirely

a photoshopped image from a civilian SSR radar, http://
i257.photobucket.com/albums/hh212/727Kiwi/Lido_
Hotel_compared_zpsd4b357a8.jpg.

We also know the Lido Hotel image was a hoax
image because it failed to show the tracks of two other
aircraft known to be in that airspace at the same time,
SIA68 and EK343: http://i257.photobucket.com/
albums/hh212/727Kiwi/MH370/UAE343Overlay_
zps49d145d2.jpg

The tracking of MH370 by Vector addition of Satellite
Burst Offset Frequency (BOF) after 18:25 UTC assumes
incorrectly that MH370 had just overflown MEKAR at
about 18:22 UTC, therefore plotted vector additions from
that location. This satellite plot track falls apart under
scrutiny. MH370 suffered a decompression catastrophe off
Vietnam after pilots had attempted to turn back. http://
i257.photobucket.com/albums/hh212/727Kiwi/MH370/
SourthernArcanotated_zpsadedf2b0.jpg.

If you want to blame anybody for the wasted effort
time and money, blame the Malaysian government for
promoting multiple hoax claims.

http://i257.photobucket.com/albums/hh212/
727Kiwi/MH370/OrlyTaitzletter_zpsac801102.jpg.
http://i257.photobucket.com/albums/hh212/727Kiwi/
OrlyTaitzletter2_zps18d9406b.jpg.

If MH370 never turned west from IGARI and flew
through the Strait of Malacca, then the only true sighting of
MH370 was by oil rig worker Mike McKay off Vietnam. If
you also accept that it crashed along the southern arc of satellite pings then it must have turned south from Vietnam like

this, http://i257.photobucket.com/albums/hh212/727Kiwi/
MH370/SourthernArcanotated_zpsadedf2b0.jpg."[184]

LOCATION MISINFORMATION

Before we get to a global game of Marco Polo and pings and
wild turns, we'd like to tell you what the chief of the search
team had to say. Get a load of this little gem:

"The search could drag on for as long as a year."[185]

What the hell is he talking about—and what exactly does
that mean? That they are going to search for a year and then
that's it, forget it, because they know it's not where they're
searching? Isn't that a bizarre thing to say? Don't you search for
something until you *find* it? Don't you search for it *not knowing*
how long it is going to take to find it? Wouldn't he hope to find
it in much *less* than a year? Why does he sound so damn pessi-
mistic about it? Doesn't that tell us something? Your car keys are
always in the last place you look for a very good reason: Because
after you find them, *that's* when you stop looking. Sounds like
the "head of the search effort" already has a game plan.

The incorrect information espoused by the authorities
regarding radar tracking and the plane's flight path has led to
what our grandmothers used to call a "wild goose chase" and,
in this case, it's a real humdinger of one too. They're not even
sure what ocean they should be looking for it in. You think
we're kidding? Just keep on reading.

In case you wondered why they are looking off the coast
of southwestern Australia for a plane that disappeared as it was
approaching Vietnam, well, it's a long story. The initial search
took place in the South China Sea, off the coasts of Cambodia
and Vietnam.[186] That's because the aircraft was on a northeasterly

heading toward Beijing, China. Extensive searches of the area at sea were fruitless, so they widened the efforts, including searching on land.[187] They searched miles of mountains and uninhabited jungles.[188] Military units, near Vietnam's borders with Laos and Cambodia, also conducted extensive searches there. But then, a report from Malaysia that their military's radar system had picked up the flight traveling west over the Straits of Malacca, changed matters dramatically. Suddenly, the search area was switched to the seas on the *western* side of Malaysia.[189]

The search was then widened, covering a huge area in the Gulf of Thailand and South China Sea around the last known position of the plane.[190] That was in addition to the searches on the western coast of Malaysia and farther northwest in the deep waters of the Andaman Sea.[191]

Then, the whole issue of the satellite pings came about and the search area was again refined:

A BBC article reads, "Investigators carrying out the search are now focusing on a refined area covering 60,000 sq km 1,800 km (1,100 miles) off the west coast of Australia. Based on the most recent analysis of satellite data, the plane is believed to have ended its journey in seas far west of the Australian city of Perth. The latest zone is some 1,000 km southwest of the area which was extensively searched with underwater surveying equipment in April. Over the last few months, the Australian authorities have been conducting an underwater depth survey of the latest search area. On 26 June, officials announced a new 60,000 sq km search area some 1,800 km west of Perth. The operation will begin in August with detailed mapping of the sea bed."[192]

It appears that the searchers haven't even been looking in the right place. The technical calculations of Gunson, in the preceding entry, reveal that the plane apparently never turned west. Therefore, when Flight 370 turned dramatically south, it was actually from a point much further east than the point used in the official calculations. That error totally corrupted the accurateness of the area designated as the search area by officials: "It is totally corrupted because they start[ed] from the wrong place," Gunson said.[193]

The whole search has been based on "the numbers," known as the "pings" between the aircraft and the Inmarsat satellites. Admittedly, Inmarsat is not infallible. CNN writer Richard Quest writes:

"When he [Inmarsat's vice president of satellite operations, Mark Dickinson] realized what had probably happened, his reaction was: 'Let's check this and let's check it again, because you want to make sure when you come to a conclusion like that, you have done the right work.'" But, Quest goes on to write, "Inmarsat and others will never say they are convinced, they are not those sorts of people. Certainty is the prerogative of those of us less rigorously trained to consider fallibility. Inmarsat's engineers will just rely on the numbers."[194]

And "get-it-wrong" they apparently have if you look at a number of ever-changing search areas where salvage teams have looked for the aircraft based on the parameters of Inmarsat's data. The "media-speak" for this misinformation is "refined search area"—a phrase we've been re-hearing for many months now.

And here's what you get when someone tries to actually summarize the whole sordid fiasco: "Although Bloomberg News said that analysis of the last satellite 'ping' received suggested a last known location approximately 1,000 miles (1,600 km) west of Perth, Western Australia, the Malaysian Prime Minister Najib Razak said on 15 March that the last signal, received at 08:11 Malaysian time, might have originated from as far north as Kazakhstan. Najib explained that the signals could not be more precisely located than to one of two possible loci: a northern locus stretching approximately from the border of Kazakhstan and Turkmenistan to northern Thailand, or a southern locus stretching from Indonesia to the southern Indian Ocean. Many of the countries on a possible northerly flight route—China, Thailand, Kazakhstan, Pakistan, and India—denied the aircraft could have entered their country's airspace, because military radar would have detected it.

"It was later confirmed that the last ACARS transmission showed nothing unusual and a normal routing all the way to Beijing. *The New York Times* reported 'senior American officials' saying on 17 March that the scheduled flight path was reprogrammed to unspecified western coordinates through the flight management system before the ACARS stopped functioning, and a new waypoint 'far off the path to Beijing' was added. Such a reprogramming would have resulted in a banked turn at a comfortable angle of around 20 degrees that would not have caused undue concern for passengers. The sudden cessation of all on-board communication led to speculation that the aircraft's disappearance may have been due to foul play."[195]

Got it? How could anybody not understand *that* simple story? If you're sitting there thinking that even a team of Philadelphia lawyers couldn't figure out those remarks, we tend to agree. In fact, a team of lawyers were probably needed to put this whole fiasco together! They cite their sources, you have to give them that much! But those words don't even sound like they're *meant* to be understood.

You can revel at the immensity of that "refined search area" on an excellent map,[196] http://upload.wikimedia.org/wikipedia/

commons/thumb/4/4e/Map_of_search_for_MH370.
png/1280px-Map_of_search_for_MH370.png.

So all those mountains of technical and sophisticated data
have only revealed to us that Flight 370 ended up "somewhere
on the red lines"—an imaginary arc based on the supposed
satellite pings, from a point high in Central Asia to a point
south of Australia.[197] Could be in Kazakhstan, could be south
of Australia. Well, well, well, how's about that for modern technology? Either they're idiots, or they're assuming that we all are.
Bet on the latter, would be our advice.

NO PHONE CALLS FROM PLANE

This issue has not been fully or properly covered in mainstream
media. There are *several* matters of relevance pertaining to phones.

Passengers' Cell Phones

The fact that no cell phone calls were made from the plane has
been "explained away" in the media:

"At 3,000 feet, you can make a call, but go much higher
and you can basically forget about that," says Wouter
Pelgrum, an assistant professor of electrical engineering at
Ohio University, in an article for *Slate*; "You don't have coverage. ... Part of the problem," he says, "is that cell tower
antennae are pointed down, toward the ground, not up into
the sky. If you're over a city, with its dense cluster of coverage,
you'll have a decent chance, but not in a rural area, and even
less so over the ocean."[198]

Some people have made the very interesting observation
that cell phone calls reportedly worked from planes at higher

altitudes during the 9/11 terror incident. It was phone calls from the passengers on those planes, you may remember, that documented the official narrative describing how some hijackers armed with "cardboard cutters" had managed to take over the entire airplane and flight crew.[199]

One explanation offered to us—a rather feeble attempt—is that the type of calls that were placed on 9/11 weren't actually from cell phones, they were from those in-flight airplane phones that one sometimes sees on the back of the seat in front of them; interestingly, *that argument failed to mention that the Boeings of 9/11 don't have those phones.*[200] So what about the United Flight 93 passengers, because most of their calls were made using GTE AirFones?[201]

Well, nice try, but you know what? Some of the calls reportedly placed from flights on 9/11 *were* from cell phones,[202] so the "rule" either applies or does not apply.

It has been posited that maybe everyone on Flight 370 was really tired and simply did not notice the several turns in the completely wrong direction that the plane made, and so didn't think they needed to make any calls to tell anyone about it. This is about as realistic as saying that 239 people all think and act in an exactly predictable manner, in other words, not realistic at all.[203] Something else was going on that prevented them from making calls.

Some people have claimed that the phone calls on 9/11 were part of an official "cover story" and never really happened. But, if the official reports of 9/11 are to be believed, then cell phone calls *were* placed and connected from flights at altitudes not thought to support cell tower functions. So, believe whichever you choose, but you only get to choose one here, folks. The point is they can't have it both ways.

Either cell phones work on planes, or they don't. If they worked on 9/11, they should have worked on 370, right? Either cell phones *can* place calls at those altitudes or they cannot. They can't conveniently work when the government explains to us how some passengers armed with box cutters managed to hijack a large commercial jetliner, but *not* work when 239 people disappear into thin air on a flight that satellites show continued airborne for seven and a half hours. And if they *can* work, then the following is a very valid question: Why were no cell phone calls placed from Flight 370? Surely everybody who had a cell phone would have tried to make a call if they could have.

In-Flight Phones

Bear in mind here that what is being referred to in those news stories about phone calls are attempted calls from *cell* phones. So what if the cell phones *didn't* work. Unless the altitude was low enough, cell phones might not have functioned to place a call or send an email.

Given this possibility, why isn't mainstream media discussing *air-to-ground* phones? The point about no phone calls typically leads to the "cell phones don't work" argument. But what about *other* phones? Malaysia Airlines offers an "air-to-ground" phone service in business class that also allows passengers to send email. But, it should be noted, the captain could shut this down if he wanted to.[204]

It defies logic to posit that no one on that plane, none of the 239 people, including flight attendants, would utilize one of those air-to-ground phones during an emergency situation if they were able to.

Ping-Ping-Ping

If a cell phone was not in, or was taken out of, airplane mode, it would send out signals every twenty seconds or so, trying to locate the nearest cell phone tower. This is why it's a good idea to put your phone on airplane mode during a flight—even if you're not worried about disrupting navigational equipment, the constant pinging will drain your battery. If a phone is on, but not on airplane mode, it might not connect to a network, but it's a source of electromagnetic radiation that can be picked up. "If you have 200 cell phones all pinging repeatedly at 6/10ths of a watt, it would be a chorus," says Paul Czarnecki, a pilot and cell phone network technician.[205]

In other words, that information would have given them the location of the airplane. According to Czarnecki, "The United States has listening stations all over the world to record and digitize every signal in the air. It blows my mind that we don't know where it is."[206] It blows our minds too.

The Call That *Did* Get Through

It should be noted that a call was also reportedly made *to the plane* from Malaysia Airlines staff. They tried to contact Flight 370 via a satellite phone call to the plane.[207] The call was reportedly unsuccessful, but information from the technical aspects of it—more pings—were said to help further refine the already multi-refined search area. What does this mean? Hard to say, when it's brought up by the media and then quickly forgotten.

Copilot's Cell Phone On

Just like passengers, pilots are supposed to shut off their cells or place them on airplane mode before pushing back from the gate. "It would be very rare in my opinion," said aviation analyst David Soucie, "to have someone with a cell phone on in the cockpit. It's never supposed to be on at all. It's part of every checklist of every airline I am familiar with."[208]

Seems pretty straightforward, right? Yet—quite contrary to that well-established and very clear policy—the copilot did indeed have his cell phone on:

"The phone of the first officer of Malaysia Airlines Flight 370 was on and made contact with a cell tower in Malaysia about the time the plane disappeared from radar, a US official told CNN ... The revelation follows reporting ... in a Malaysian newspaper that the first officer had tried to make a telephone call while the plane was in flight."[209]

Get a load of the major implications of that point! Doesn't it bring up some questions? The article goes on to state: **"The details do appear to reaffirm suggestions based on radar and satellite data that the plane was off course and was probably flying low enough to obtain a signal from a cell tower, the US official said."[210]**

The "details do appear" to do that—those nasty details. Leading to the question:

So why were there no calls?

STOLEN PASSPORTS

Two passengers on Flight 370 were found to be traveling with false documents, passports that were stolen and did not match their actual identities. Those two passengers were from Iran. Initially, they were prime suspects as to a possible cause for the flight's disappearance. The specifics of their case are covered in Chapter Ten.

FAKED PHOTOGRAPHS RELEASED TO MEDIA

As amazing as it sounds, Malaysian authorities <u>admitted that they released a faked photograph of the two Iranian passengers</u> on the flight who were accused in the press of being hijackers. They said it was a harmless mistake. It's an obvious falsification. Check it out in the photo section of this book. Again, we're not making any of this stuff up, folks. It's yet another case of reality being stranger than fiction:

> Fears of a cover-up over the fate of flight MH370 grew yesterday after claims that a photo of two passengers was tampered with. Images of two men who boarded the Malaysia Airlines jet with stolen passports appear to show them having the same set of legs. CCTV [closed circuit cameras] footage stills released by officials three days after the Boeing 777-200 vanished from the skies shows the pair with identical green trousers and brown shoes.[211]

Actually, it's a bit worse than that. The lighting is identical, the positions are identical, even the shadows are identical—a physical impossibility.

A spokesperson for the Malaysian police admitted the image of one man had been placed on top of the other when they were photocopied. But they said it was not "done with malice or to mislead."[212] *Right*. That makes sense. The whole controversy was apparently started when a Twitter user noticed the "sameness" of the legs of the two alleged hijackers (as they were already being called) and tweeted, "They both have the same legs, edited or coincidence? And you guys believe our gov is not hiding anything."[213]

After the jig was up and it started to blow up in their faces, the police politely explained that it was just a simple mistake by someone at the photocopier that day. *O*-key Dokey. Whatever you say.

PILOT'S FLIGHT SIMULATOR

While not necessarily an anomaly, most people considered this one unusual. When police conducted a search of the pilot's home, they found a flight simulator there. What was considered interesting about that, especially to investigators, was the fact that they were set for five airstrips in the Indian Ocean, even though Flight 370 was not supposed to be going anywhere near it.[214] The five locations had airstrips that were sufficient, but difficult, for a Boeing 777—runways of approximately 1000 meters. Three of those runways were in India and Sri Lanka; one was in the Maldives (where many eyewitnesses reported a low-flying commercial airliner); and one was at the secret US military base at Diego Garcia, which is on a small island south of India.

Since Malaysian officials stated that whatever happened to Flight 370 was the result of a "deliberate act," attention obviously was directed at the passengers and flight crew.

An article in *The Malaysian Mail* reads, "The police had seized the flight simulator from the 53-year-old pilot's house in Shah Alam on Saturday and reassembled it at the police headquarters where experts are conducting checks. The Transport Ministry has said that the police also searched the home of Zaharie's copilot, Fariq Abdul Hamid, on the same day. Also on Saturday, Prime Minister Datuk Seri Najib Razak said MH370 was diverted deliberately after someone on board switched off the Boeing 777's communications systems. He said investigations were now being refocused at the crew and passengers aboard the plane."

Since it seemed odd for a commercial pilot to be practicing a landing on those remote airstrips, it raised some concerns. The Malaysian authorities obviously found it of importance. A source for them said, "We are not discounting the possibility that the plane landed on a runway that might not be heavily monitored, in addition to the theories that the plane landed on sea, in the hills, or in an open space."[215]

And, if you're wondering why there were conspiracy rumors about Flight 370, ask the authorities, because they acknowledged that possibility; "Although Defense Minister Datuk Seri Hishammuddin Hussein denied yesterday that the plane had landed at US military base Diego Garcia, the source told the daily that this possibility will still be investigated based on the data found in Zaharie's flight simulator software."[216]

On the *other* hand, if you're questioning whether it's strange for a pilot to have a flight simulator he or she practices on at home, it's not. In fact, it might simply mean that the person is a good pilot. Here's how one pilot responded to that, "Is it normal for a pilot to have a flight simulator at home? It's not that uncommon; pilots generally love their job. I've got a flight simulator program on my computer, but rarely use it. Some pilots love their job so much that in their spare time they like to fly simulators, fly aerobatics, and train other pilots. The job is more of a lifestyle than a 9:00–5:00 job, so I don't see it as concerning that the captain had set up a very basic flight simulator at home."[217]

PASSENGER'S PRE-FLIGHT CONCERNS

In an event that deserves mention but does not appear to us to be anything sinister, a passenger of Flight 370 was apparently very disturbed about his possibility of not surviving, because he specifically left his wife his watch and wedding ring "in case anything happened" to him. His wife said that his words to her were, "If something should happen to me then the wedding ring should go to the first son that gets married and then the watch to the second."[218]

His wife thought it was worth mention—and we honor her wish. He also took a lot of family photos. *But*, he was also going off to a new job in Mongolia, so, as we said, that one seems quite explainable.

PASSENGER'S REPORTED MESSAGE

Philip Wood is listed as a passenger on Flight 370 on the official "MH370 Passenger Manifest" released by Malaysia Airlines. He is listed as Philip Wood, American, Age 51.[219]

Wood was a technical storage executive for IBM-Malaysia— you can search and find his LinkedIn page—and he was a frequent business traveler who had an iPhone 5. It's been reported that he sent a message after Flight 370 was hijacked, via his cell phone. There is a great deal of "debate" about the message, due to the seriousness of the implications. Some say this story is just an elaborate hoax. If it *is* true, it's a very big deal.

Jim Stone, a blogger who says that he is a former "high-level employee of the NSA in a position much higher than Edward Snowden,"[220] runs a blog, jimstonefreelance.com, that is popular in alternative media, and on his blog maintains that the message is authentic. To read his theory in full, visit http://jimstonefreelance.com/phillipwood.html.

If you prefer a short summary, some researchers believe that Philip Wood sent a message from his iPhone and attempted to send a photo after disembarking from Flight 370. The photo was not received, but, allegedly, the encrypted information embedded in the message revealed the GPS location of the cell phone at the time the message was sent—and the coordinates are a match with Diego Garcia, the remote secret US military base in the Indian Ocean. It's said that the text message was as follows:

> I have been held hostage by unknown military personal after my flight was hijacked (blindfolded). I work for IBM and I have managed to hide my cellphone in my ass during the hijack. I have been separated from the rest of the passengers and I am in a cell. My name is Philip Wood. I think I have been drugged as well and cannot think clearly.[221]

Although we won't get into it here, Stone's analysis of the software infrastructure for embedding on the iPhone 5 lends substantial credence to the GPS coordinates of the message as being correct. As researchers are also wont to point out, the fact that voice recognition software would print "personnel" as "personal" is further indication of its authenticity.[222]

Wood's girlfriend, Sarah Bajc, has made a very public and commendable effort to learn the truth about Flight 370. You can listen to her in several videos online, discussing—quite logically— why she is skeptical about what we have been told. There are some good videos of this if you visit the URL, www.msnbc.com/ ronan-farrow/watch/plane-passengers-partner-still-skeptical-209483331906; you could also simply Google "Sarah Bajc."

Since Flight 370 has "disappeared," Bajc, who was preparing to move with Wood to Beijing, has received death threats and creepy warning messages on her cell phone, such as "I'm going to come and kill you next."[223] She's also stated that a source revealed to her that Flight 370 had a military plane escorting it.[224] Her conclusion, after many months of searching for answers, is that the search was "too incompetent to be accidental."[225]

Among other grievances, Bajc has criticized the Australian government's role in the investigation after they stepped in to lead the Indian Ocean search. "She said the Australians should be 'embarrassed' for targeting a dead-end search based on underwater pings, and for not forcing the Malaysian government to release all of their records."[226]

Since the search has found no clues as to what happened to the plane, even in this age of high technology, Bajc believes that, for some reason we don't currently know, the investigation was sabotaged. She said, "I think that if the existing investigation team is left in charge … we may not ever find the plane because I believe there are active steps being taken to interfere with finding the plane." Bajc says she isn't sure how much is being covered up, but knows something is being concealed. "Failure to release information—whether it's obfuscation, you know, actually covering something up—or dishonesty … creating false evidence or just hiding something, right?"[227]

Bajc's studied observations echo those of the former Malaysian Prime Minister. "We don't know why or what is being covered up, but something is being covered up," she said."[228]

NO PLAUSIBLE EXPLANATION

As you have no doubt noticed, the authorities *still haven't offered a plausible explanation* of what happened to Flight 370, or even a believable accounting of the actions of the authorities during the event.

BOEING HONEYWELL UNINTERRUPTIBLE AUTOPILOT

And now, with only slight further ado, ladies and gentlemen, for the award of the Crème de la Crème of Controversy in the field of anomalies, we announce BHUA—the big acronym for four not-so-little words, Boeing Honeywell Uninterruptible Autopilot.

Advanced avionics have reached the point where, *reportedly*, though disputed, it is quite possible, even sometimes desirable, to take over the controls of a large aircraft from the pilot *remotely*, such as in a hijack situation. An entry on *21st Century Wire* reads, "The Boeing 777, along with other Boeing models, can in fact be flown remotely through the use of independent embedded software and satellite communication. Once this advanced system is engaged, it can disallow any pilot or potential hijacker from controlling a plane, as the rooted setup uses digital signals that communicate with air traffic control, satellite links, and other government entities for the remainder of a flight's journey. This technology is known as the Boeing Honeywell 'Uninterruptible' Autopilot System."[229] It's patented, and it apparently would be public knowledge, except for the fact that they don't *want* us to know about it. So if there's anything about it at all in mainstream media, it's things to the effect of "no, no, no, that's just a weird conspiracy theory that's never been proven." But we're not talking about conspiracy theories—we're talking about a patented technology that actually seems to exist. We'll explore this, and other issues like it, later on.

ADMISSIONS FROM PUBLIC OFFICIALS

Advanced avionics, such as the type in the entry immediately preceding this one, were operational on Flight 370; so we are told by the former Prime Minister of Malaysia, who publicly claimed that the plane was equipped with those special avionics that would indeed allow it to be controlled by others. He was very specific, stating that "the plane was equipped so that it could be taken over 'remotely by radio or satellite links by government agencies like the Central Intelligence Agency.'"[230] It's not *us* saying that—it's the guy who was the freaking prime minister of that country. *And* he specifically named the CIA and Boeing as the culprits who should be explaining to us what *really* happened, because they know.[231]

Adding to the air of already anxious circumstances are other odd statements—that seem more like admissions—from high-placed public officials. On September 12, 2014, more than six months after the disappearance of all on board Flight 370, the chief of the national police of Indonesia, General Pol. Sutarman, laid bare an amazing revelation. He stated that he had spoken to the chief of the Malaysian national police and was told by him that it was precisely known what had happened to Flight 370. Here are his exact words, which several people witnessed:

"I spoke to the Malaysian Police Chief Tun Mohammed Hanif Omar. I actually know what had actually happened with MH370."[232]

He actually said that, and it was at a large meeting between an aircraft company and public officials, so there were many witnesses. According to a *Mirror* article, "General Sutarman's comments were witnessed by a number of high-ranking police officers and representatives of Lion Air [one of Indonesia's largest airlines]."[233]

As one might imagine after seeing that statement, if not before, many people have assumed that the governments involved have known the truth about Flight 370 right from the day it "went missing," which would certainly seem to make sense. They just aren't planning on telling *us* the truth, apparently. But high-ranking officials have clearly implied that it *is known* what happened to the flight. It's just not being publicly revealed. Keep in mind that this revelation was in addition to the aforementioned statement, by no less than the former prime minister of Malaysia, that someone is hiding something.

As further testament to the veracity of these statements, General Sutarman's comments were quickly dismissed and not disseminated with any sense of importance in mainstream media. (Gee, *there's* a surprise.) "The Inspector General of Police in Malaysia, Khalid Abu Bakar, has told of his shock over the comments and said he would speak to the Indonesian police chief."[234] We bet he did. And if the good general wants to keep his job, then he'd better be a good boy and keep quiet now too. So far he has.

The immediate denials and the fact that the good general was quickly muzzled from making further public statements that might embarrass governments and their allies, are further indication that the truth about Flight 370 is known, but being covered up. It's the typical pattern of a governmental cover-up and, in this case, one that appears to be international in scope.

Nine

Comparison Between Malaysia Flight 370 and Other Recent Crashes

There is nothing in aviation history *even close* to the saga of Flight 370. And if you believe reports to the contrary—that there *are* indeed similar cases—then you have been hoodwinked by mainstream media and should play close attention to how the media has blatantly misrepresented the events of other flights, such as Air France 447, covered in the following pages.

As you will see from the following comparisons, there is literally *nothing* in the entire history of aviation that compares credibly with what we are *told* are the facts about Flight 370. The notion, popularized in mainstream media, that this sort of thing "happens"—we *do* lose planes—is patently false. This "sort of thing" has *never* happened. There has been nothing remotely like MH370 in the modern era.[235]

As Dr. F. Robert van der Linden, Chairman of the Aeronautics Division of the National Air and Space Museum in Washington, DC, expressed it, "*For an airliner—just to disappear? That's unheard of.*"[236]

CRASH INTO WATER—AIR FRANCE FLIGHT 447—INTO THE ATLANTIC OCEAN NEAR BRAZIL—JUNE 1, 2009

A flight that is often compared to 370 is Air France Flight 447, also a large commercial jetliner—an Airbus A330 with three pilots—that temporarily "disappeared" over the Atlantic Ocean in 2009.

You may have heard, in mainstream media, the *quite* incorrect statement that it "took two years to find Air France Flight 447" after it "disappeared" when it crashed into the ocean. That has been offered as a weak excuse for the complete absence of information regarding the whereabouts of Flight 370. <u>That information is wrong</u>. It did *not* take two years to find the plane at the bottom of the ocean. It took two years to *complete the salvage operation* of the plane from the bottom of the ocean. That's a huge difference. They knew almost *exactly* where the plane went down. In fact, they found wreckage floating at the crash site, the *day after* it went down. *It was not lost*, except for a few hours. But, being in deep waters, it was a very complicated salvage operation.

However, it's extremely misleading to compare Air France Flight 447 to Flight 370, particularly considering the fact that the authorities reportedly are not even sure what ocean 370 went down in, *if* it's an ocean. In fact, the way that CNN misleads the reader regarding the experience of Flight 447 is outrageously wrong. In an article published just a few days after Flight 370 went missing, and in which they compared it to Flight 447, among others, they prefaced the article with the following remarks: "While such situations are rare, the puzzling disappearance of the Malaysian jetliner is not the first time a plane has vanished without a trace. Here are nine cases of mysterious plane disappearances and disasters. Some remain unsolved, decades later."[237] That implication is far from accurate. They found the crash site for Flight 447 *the day after it crashed*, along with a three-mile wide debris field.[238] Just don't tell CNN that because, apparently, it would ruin the slant of their story. So, please pay close attention to the examples in this chapter because they will correct a lot of information that is extremely misleading.

On June 2, 2009, just one day after the crash of Flight 447, the Brazilian Air Force spotted a huge oil spill and wreckage in a debris field of jet fuel and random pieces of the aircraft that was a full three miles long.[239] They recovered an aircraft seat, an orange life vest, a drum, and kerosene and oil.[240] By June 6, they had recovered two bodies, a blue seat, a briefcase, and a backpack containing a computer.[241] The following day, they recovered the first major piece of wreckage—the plane's vertical stabilizer. Then began the long process of lifting up parts from the ocean bottom.

To compare the crash of Flight 447 to what we have heard pronounced as "the verdict" of Flight 370—that it crashed into the Indian Ocean—consider the following about Flight 447: The aircraft broke apart upon impact with the ocean.[242] Debris was easily visible and spread over a three-mile field, including a very visible oil slick. So take a good long look at Flight 447. It's a great example of how planes don't just disappear.

Even more noteworthy regarding the loss of Flight 447 is that even though it crashed into the ocean, its flight systems sent valuable information, including twenty-four maintenance messages, to its airline, Air France, that told the airline what had happened to the plane and where and when it happened.[243] So that raises some obvious questions: Shouldn't Malaysia Airlines or Boeing know *exactly* what happened to Flight 370? Doesn't it seem quite naive to suggest that one person, turning off one switch in the cockpit, if that even happened, totally eliminates all communications with one of the most sophisticated aircraft models in existence?

ELECTRICAL FIRE—SWISSAIR FLIGHT 111—SEPTEMBER 2, 1998—ATLANTIC OCEAN NEAR NOVA SCOTIA

Another flight compared to 370 is Swissair Flight 111. This flight, like 447, is often cited as an example of how difficult it is to recover planes that crash into the ocean. In other words, the authorities use them as examples to support why they haven't yet found Flight 370—they say it's at the bottom of an ocean, so it's extremely hard to do so. However, again in this case, that claim is extremely misleading. It's true that it took two years to recover the wreckage of Flight 111, but that was to *complete* the recovery procedures, and *they knew where the wreckage was.*

Air traffic control knew the *exact* coordinates of the crash. The crash location was determined to be at global coordinates of 44°24′33″ N 63°58′25″ W, with a margin of error of only three hundred meters.[244] And with those coordinates in mind, the authorities found the plane quickly and rather easily. The aircraft was located at ocean bottom by the submarine HMCS Okanagan, via sonar detection of the underwater beacon signals. Both beacons were quickly retrieved by divers from the Canadian Navy: the flight data recorder on September 6, 1998—just *four days* after the crash; the cockpit voice recorder on September 11, 1998—just nine days after the crash.[245] In other words, the episode was *not at all* similar to the events of the search for Flight 370. The authorities knew *exactly*, and *quickly*, what happened to Flight 111: At exactly 22:31 AT (01:31 UTC), the plane struck the Atlantic Ocean at a speed of 345 mph and with a g-force of approximately 350, "causing the aircraft to disintegrate in less than a second."[246] That impact caused the plane to break into, literally, *millions of pieces.*[247] Much of the debris sank to the ocean floor, however, some debris was found floating in the crash area, and other debris washed up on the nearby shorelines over the following weeks.[248]

Part of the reason air traffic control knew exactly what had happened and where the flight went down was because there *were communications during the in-flight emergency.*[249] The entire transcript of those conversations is even publicly available online; no one has tried to hide them. They're very dramatic to read, especially in light of the fact that every person on board that plane died a short time after those words were spoken, http://aviation-safety.net/investigation/cvr/transcripts/atc_sr111.php.

Quite unlike Flight 370, *much was known* about Swissair 111, and it was known quickly. In other words, any comparison you can make with it and Flight 370 is virtually meaningless.

LOST IN BAD WEATHER—AIR ALGÉRIE FLIGHT 5017— JULY 24, 2014—MALI, WEST AFRICA

A flight that *does* warrant comparison to Flight 370 is Air Algérie Flight 5017. The plane had taken off in Burkina Faso, a nation in northwestern Africa, en route to Algiers, Algeria, the African nation to its north; the route took it over Mali. Contact with the flight was lost about fifty minutes after its takeoff, and the plane disappeared from radar. Although it went missing and was lost in very bad weather conditions, the crash site was found within "several hours" after it lost contact with air traffic control,[250] even though the plane was entirely "disintegrated."[251]

Again, everything was contrary to the case of Flight 370. With Flight 5017, minute details were known about precisely what had happened to it, as well as precisely when they had happened. Satellite images provided "high-resolution infrared and day/night band imagery along the flight path" that even identified the aircraft impacting the ground, *through* the infrared images of the storm.[252]

So, it would seem that the reality of the situation is that satellite imaging *tracks everything*. Everything, that is, except Flight 370.

SHOOT-DOWN—MALAYSIA AIRLINES FLIGHT 17—JULY 17, 2014—WENT DOWN OVER UKRAINE

In an occurrence so bizarre that people thought they must be hearing wrong, *another* Boeing 777-200ER, *also* from Malaysia Airlines, went down over the Ukraine only four months after Flight 370 "disappeared." It was such an odd thing that most people at first actually believed that they were hearing about Flight 370—but people were giving the wrong flight number.

It was just too surreal. That shoot-down resulted in a huge debris field. Reporters for the *Wall Street Journal* wrote, "Debris spotted on the ground appears to be spread over at least six miles and across three Ukrainian villages."[253]

That's a *lot* of debris. But the *most* dramatic differences between the two Malaysian triple-sevens was not what happened to them; it was what happened immediately *after* the incidents. When Flight 17 went down, the authorities—specifically *including* US intelligence officials—presumed to know just about everything about it, on an immediate basis, and revealed their vast knowledge about what had transpired very quickly to the world. US intelligence immediately revealed that info, as well as the fact that pro-Russian troops in Ukraine had access to that specific weapon.

US intelligence got very busy—even on Facebook, folks—and announced that radar clearly tracked the launch of a missile from an SA-11 launcher that hit Flight 17 over a portion of the Ukraine where Russian rebels were active. It was revealed that US intelligence even *tracked the specific missile* that was launched at it.[254] That's pretty darn specific. So, apparently everybody's radar was working pretty well *that* day, huh, kids? Doesn't look like a lot of uncertainty *there*, does it? For Flight 17, they even knew what specific type of ground-to-air missile brought the plane down.

A CBS article reads, "The evidence presented by senior intelligence officials consists in part of pictures posted on social media. They show an SA-11 surface-to-air missile launcher, also known as a 'buk,' driving through rebel-controlled areas of Ukraine in the days before the shoot-down."[255] They even went so far as to announce how exactly the missile launcher got there and also released, what one would think would be coveted intelligence, that they had confirmed the authenticity of "phone calls intercepted by Ukrainian intelligence in which rebels are talking about positioning the SA-11."[256]

It was disclosed that "advanced spy satellites" had detected the explosion of Flight 17, via satellite data from infrared sensors.[257] They were even able to *pinpoint* the location;

"American intelligence agencies concluded that the Boeing 777-200 was struck by a Russian-made SA-11 missile fired from a rebel-controlled area near the border in Ukraine ... an area near the small towns of Snizhne and Torez, about midway between the rebel strongholds of Donetsk and Luhansk."[258]

They knew it *all*, Charlie Brown. When it fits their geopolitical motivations, their technology all seems to work just fine. Flight 370 was exactly the opposite. *Nobody seems to know anything particularly helpful,* or at least be admitting it. They're not even sure what ocean it might be in, if it really is in one! But Flight 17—and, apparently, *everything around it*—was easily tracked by the same sophisticated technologies that supposedly failed for Flight 370. Isn't that just a little bit too dichotomous for common sense? Two different aviation disasters involving the exact same type of plane, and only months apart. How can they know everything about one of them, yet claim to know virtually nothing about the other?

UNRESPONSIVE PILOT DUE TO DEPRESSURIZATION— PRIVATE PLANE—SEPTEMBER 5, 2014—INTO THE SEA, NEAR JAMAICA

Another case of an unresponsive pilot—or "ghost flight"— was one that also garnered a lot of attention recently. A high-performance passenger plane of a high-profile real estate developer, Laurence Glazer, and his wife, Jane, was en route from Rochester, New York to Naples, Florida. The plane was a single engine turboprop Socata TBM700.[259]

The aircraft apparently started to depressurize not long after takeoff. The air traffic control audio recordings reveal that the pilot radioed ATC: "We need to descend down to about [18,000 feet]. We have an indication that's not correct in the plane."[260] The pilot did not declare an emergency and air traffic control told the pilot to fly at 25,000 feet. The pilot responded "We need to get lower." ATC said "Working on that."[261] The pilot was then cleared to a level of 20,000 feet, but it was too late. Apparently they had been overtaken by the effects of the depressurization. The pilot never responded and the aircraft maintained a flight level of 25,000 feet. The plane just kept flying along on autopilot.[262]

US Air Force military fighter jets were launched from Florida and South Carolina to intercept the flight and they soon rendezvoused with it, still at 25,000 feet.[263] Under the direction of NORAD (North American Aerospace Defense Command), two F-16 aircrafts were launched from McEntire Joint National Guard Base in South Carolina and monitored the plane's flight.[264] As the flight progressed south, monitoring was handed off to two F-15 fighter jets that were launched from Homestead Air Reserve Base in Florida.[265] The fighter jets got close enough to the passenger plane to look right into it and see the pilot slumped over in the cockpit. The fighter pilots got so close that they could even determine that the pilot was still breathing, though he appeared unconscious.[266] They also noted that the windows of the small plane had begun frosting over, an indication of a loss of cabin pressure.[267]

The two F-15s "escorted the aircraft until it entered Cuban airspace. At this time, they broke off with the possibility of reconnecting with the aircraft over international waters south of the island."[268] The small plane eventually ran out of fuel and crashed into the ocean, near Jamaica; the two F-15s were not present at the time because they had to return to base to refuel.[269] In retrospect, the pilot should have immediately declared an emergency, bypassing the need for permission to begin the necessary descent, because once the plane reaches an altitude level of 10,000 ft., the air is again breathable.[270]

Even though it was a smaller passenger plane, a comparison to Flight 370 is valid because the theory of depressurization and an ensuing "ghost flight" is the one that most experts seem to have "settled on" as what happened to Flight 370. A better example of this would be the Helios Airways Flight 522. It also depressurized in flight, but was a large commercial airliner like Flight 370.

UNRESPONSIVE PILOT DUE TO DEPRESSURIZATION—HELIOS AIRWAYS FLIGHT 522—AUGUST 14, 2005—GREECE

Another flight emergency comparable to Flight 370, at least at this stage in the investigation, is that of Helios 522. Helios 522 was a Boeing 737 aircraft headed from Cyprus to Athens, and its flight was the deadliest in Greek aviation history. The plane depressurized in flight, causing a lack of oxygen that, like with Laurence Glazer's private plane, led to pilot incapacitation, and that eventually brought the plane down. Like with Glazer's plane, Helios 522 continued flying for a while with the crew incapacitated, also labeling it a "ghost flight."

With Helios 522, warning lights came on in the cockpit and the flight crew made several communications back and forth with ground control, even taking questions from the ground engineers, the maintenance personnel in charge of the pressurization systems.[271] After those communications, the situation unresolved, the flight crew apparently lost consciousness, and failed to respond to many transmission attempts from air traffic control. As the plane approached the large Athens airport, it automatically went into a circular holding pattern, via the plane's computerized flight management system and automatic pilot, and it then remained in that pattern.[272]

Fighter planes—two F-16 military combat aircrafts—were quickly sent up to intercept the aircraft. The cockpit and cabin of the airliner were clearly observed by the fighter pilots; they could even see the oxygen masks dangling in the air. The captain's seat was empty. The copilot was slumped over the controls, apparently unconscious.[273]

Things then got even *more* dramatic, as a flight attendant on the plane, Andreas Prodromau, entered the cockpit and took over the captain's seat. Prodromau held a United Kingdom commercial pilot's license; he was not licensed to fly a large airliner, but he knew how to fly a plane. Prodromau, unlike everyone else on the plane, had survived by an ingenious action. He went from mask to mask in the passenger cabin of the depressurized plane, until he worked his way to an emergency portable oxygen tank at the front, which he then switched to in order to continue breathing. Everyone else's oxygen ran out as there is ordinarily a supply lasting only fifteen minutes; that is usually plenty of time to get the aircraft down to a flight level of 10,000 ft. in a depressurization situation where, at that level, oxygen supplementation is no longer necessary.[274]

It must have become apparent to Prodromau that there was a crisis in the cockpit. He at first struggled to gain access to the cockpit of the plane because of the levels of security on the special door, but he succeeded, presumably by entering the proper security access code.[275] But, shortly after he entered the cockpit, the left engine flamed out from a loss of power and about ten minutes later the right engine of the plane also failed. He was never able to gain control of the plane or even to communicate with air traffic control or the two fighter pilots.[276] It was too late for effective action to be taken and the aircraft crashed into a hill outside the airport. There were no survivors.[277] Quite possibly, the catastrophe could have been avoided if the flight crew simply had the pressurization control lever where it was supposed to be, on automatic. That was the question that the ground engineer was asking the crew, and one which he never got an answer to: "Can you confirm that the pressurization panel is set to AUTO?"[278] It was actually set to "MANUAL" with the aft outflow valve partially open.[279]

Depressurization has been cited as a possible explanation to what is most frequently theorized to have happened to Flight 370, a crash. However, note the numerous dissimilarities between 370 and the Helios flight:

1. The pilot of Helios 522 contacted air traffic control about a warning light in the cockpit. The pilot of Flight 370 did not.
2. The crew of Helios 522 had extensive communications with ATC regarding how to resolve the problem. The crew of 370 did not.
3. Fighter planes were quickly sent up to look into the cockpit of Helios 522 and determine the situation; for 370—according to the official report—none were sent.

4. The authorities knew exactly where Helios 522 crashed. For 370 they, *reportedly*, do not.

5. The authorities have recovered the wreckage of Helios 522; for 370 they have not.

"MIRACLE ON THE HUDSON"—US AIRWAYS FLIGHT 1549—JANUARY 15, 2009—NEW YORK

We've already mentioned this flight and most Americans are familiar with it. It was the dramatic landing of a large airliner on the Hudson River in New York City in 2009. After takeoff from LaGuardia Airport, just 4.5 miles out and still during its initial climb, the Airbus A320 hit a flock of Canadian geese, knocking out the aircraft's engine power.[280] At an altitude of 2,800 feet, in a steep climb, the windshield of the aircraft turned brown from geese strikes and the engine-hit started slowing the plane down as it continued increasing in altitude. The very experienced Captain Chesley "Sully" Sullenberger took the wheel manually and contacted air traffic control: "Hit birds," he said. "We've lost thrust on both engines. We're turning back towards LaGuardia."

The engines of the plane flamed and banged and then shut down in a very ominous silence, as the smell of jet fuel, or burning birds, filled the cabin.[281] Captain Sullenberger was immediately cleared for landing back at LaGuardia, but by the time he was informed of this, it was too late. He told air traffic control that he would not be able to make it that far, due to the plane's low altitude, and asked if it was possible for an emergency landing at Teterboro Airport in New Jersey, which was closer to where he was than LaGuardia.[282] Air traffic control obtained permission from Teterboro and relayed that to the flight crew; but by then, they realized they would not have sufficient altitude to make that either. The crew sent the dramatic final transmission to air traffic control, "We can't do it. We're going to end up in the Hudson."[283] Controllers could actually see the plane, descending dramatically, less than nine hundred feet over the George Washington Bridge. Over the plane's intercom, the captain announced "Brace for impact" and set it down on the river, in a smooth, unpowered ditch [a controlled landing].[284] All were evacuated, and there was no loss of life.

Quite fortunately, the flight crew reacted almost exactly the way they were trained to do in an emergency situation. But that's the larger point. Strangely enough, the public tends to look at that event from an incorrect perspective. Captain Sullenberger—and almost everyone associated with the flight's smooth landing— were revered as heroes in the press and given a king's welcome, complete with awards and adorations. Author and aviator Patrick Smith puts that all into the proper perspective:

"Together with the majority of my colleagues, I have the utmost respect for Captain Sullenberger. But that's just it: respect. It's not adoration or a false, media-fattened misunderstanding of what he and his crew faced that day. As the public has come to understand it, Sully saved the lives of everybody on board through nerves of steel and superhuman flying skills. The truth isn't quite so romantic."[285]

As far as the ditch landing, it was done under pretty ideal conditions. The flight had just taken off and was still very close to the airport, the water was smooth as glass, visibility was good, the plane was under control; everything was perfect for a controlled ditch. Patrick Smith pointed out the good fortune:

"And nowhere in the public discussion has the role of luck been adequately acknowledged. Specifically, the time and place where things went wrong. As it happened, it was daylight and the weather was reasonably good; there off Sullenberger's left side was a 12-mile runway of smoothly flowing river, within swimming distance of the country's largest city and its flotilla of rescue craft."[286]

The same author pulled no punches when it came to a matter of calculating what would have happened *without* such lucky conditions:

"Had the bird-strike occurred over a different part of the city, at a lower altitude (beyond gliding distance to the Hudson), or under more inclement weather conditions, the result was going to be an all-out catastrophe, and no amount of talent or skill was going to matter."[287] Even though Flight 1549 had a perfect landing, the hard impact with the water still tore large holes in the plane's fuselage.[288] Can you imagine if that had been in the Indian Ocean? As a documentary produced by the Smithsonian put it when comparing Flight 1549 to Flight 370: "But that was on a fairly placid river—not an ocean … it would be at the mercy of a sea where wind speeds reach forty miles an hour and the waves can be almost twenty feet high."[289]

Also note that the pilot immediately notified air traffic control that the crew was declaring an emergency.[290] Unlike many of the other referenced flights, there was no emergency declared by anyone on Flight 370, and no one has been able to explain why.

FUEL EXHAUSTION DUE TO HIJACKING—ETHIOPIAN AIRLINES FLIGHT 961—INDIAN OCEAN—NOVEMBER 23, 1996

Ethiopian Airlines Flight 961 was a Boeing 767-260ER that was hijacked as it flew out of Ethiopia. If you would like to see what a "controlled ditch" at sea *really* looks like, there are amazing film clips on YouTube of its landing. In fact, they might be a little bit *too* exciting. See for yourself; you can either search online for "Ethiopia Flight 961 crash" or enter the links www.youtube.com/watch?v=AvtYtvd5x60 and/or www.youtube.com/watch?v=SnBkH_UuxcI.

Of the 125 people on Flight 961, only fifty survived the crash. Most of the fatalities were due to drowning and were caused because the passengers inflated their life vests *inside* the plane and then became trapped when it quickly filled with water after the crash landing. Had they waited and inflated their vests after they evacuated—just like in the instructions on that pre-flight video that most of us have viewed—most of them would have survived.[291] It's a sad reality, but one everyone should keep in mind in case of an emergency.

This video shows the plane going down—and quickly. Inside, the pilot, out of fuel and falling short of the airport, was attempting to put the plane down parallel with the beach. Bear in mind as you watch, that this was under relatively good conditions. Visibility was clear, the waves were low, and the pilot even ensured that he was in line with the waves, instead of coming in against them, in an attempt to lessen the impact.[292] But it's still just too much plane hitting too much water.

The point is a huge one: A passenger jet has *never* been safely landed on the open ocean.[293] And the debris from an aircraft impacting ocean waters has always been substantial.

After you've seen what happened when Ethiopia Flight 961 crashed, the math is pretty simple: If Flight 370 had to be ditched in the middle of the Indian Ocean, especially if it was at night, the odds are that it fared even worse than Flight 961. The impact would likely have devastated the structure of the aircraft, leaving a huge field of debris, much of which would float and, presumably, wash up ashore somewhere within a period of days or weeks.

HIJACKING—ETHIOPIAN AIRLINES FLIGHT 702—FEBRUARY 17, 2014—HIJACKED OVER SUDAN

Another Ethiopian Airlines Boeing 767 was hijacked; this time in February 2014. Thankfully, the flight landed safely, and it's interesting to note how sophisticated and astute the radar tracking of the plane was. The plane was tracked by radar every minute of its flight. On YouTube, you can watch what the tracking looked like to air traffic control as you hear the actual air traffic recordings. The personnel even negotiate as they communicate, so you can hear how an air hijacking is really handled: in a very calm, professional, and thorough manner.[294] It's dramatic stuff. The drama is also detailed in the text beneath the video, just click "Show More." Check that out at www.youtube.com/watch?v=k5gnwts6IMU.

Ten

Explanation #1—Weather

According to reports, weather conditions on the route Flight 370 was supposed to travel would not have been problematic for flight navigation in the early morning hours of March 8, 2014, or during the hours following. An ABC article reads:

"No distress signals were received before the plane disappeared, and there were no reports of bad weather. The fact that the plane's systems were gradually switched off would seem to rule out a sudden catastrophic mid-air incident."[295]

If anything, the weather was "a plus," which would have been of assistance to the flight crew for navigational purposes in the event of an electrical failure, which has been put forth as a strong possibility.

Keep in mind, also, that there was moonlight that night, and moonlight has a bearing in regard to witness sightings. It would help an observer because it would increase visibility and the person's ability to sight an aircraft in the night sky. Think back to earlier in the book when we discussed moonlight in regard to it aiding people who believed they saw Flight 370 after it was "lost."

Weather can effectively be eliminated as a causative factor in the disappearance of Flight 370.

Eleven

Explanation #2—Hijacking

The initial indication from Malaysian authorities as to what happened to Flight 370 was that the flight was hijacked. The fact that the transponder was reportedly turned off was a gigantic red flag that the flight crew may have been compromised and under physical duress by hijackers. Shortly after the flight went missing, Prime Minister Razak of Malaysia made a public announcement that was an important declaration concerning "deliberate action." On March 15, 2014, seven days after the plane went missing, Razak said investigators believed someone on board had deliberately turned off its communications systems and diverted it well west of its planned flight path. He said the plane's systems were gradually switched off and it then flew westward over Malaysia before turning to the north-west. He said such actions were "consistent with deliberate action by someone on the plane." New data showed the last communication between the missing plane and satellites at 8:11 A.M. Malaysian time—almost seven hours after it dropped off civilian air traffic control screens.[296]

It was also determined that the plane's primary tracking device, the transponder, "was disabled before" the final message received from the flight crew.[297] That final message was a calm and normal "Good night, Malaysian Three Seven Zero," as the flight crew bid goodbye to Kuala Lumpur air traffic control, which had just handed them off to Vietnam air traffic control at precisely 1:19:29 A.M.[298] The act of turning off a transponder—if it was turned off—is seen as having some big ramifications. John Goglia, a former US National Transportation Safety Board member, said:

"Given that this airplane has so many redundant electrical systems on it, my first reaction would be that somebody ... purposely turned it off. A pilot would not do that. Somebody that didn't want to be seen very well would do that."[299]

Other aviation experts concurred with that assessment. John Ransom, a retired commercial pilot and safety consultant, said the situation is suspicious. "This was a fairly modern airplane with a bunch of capability to communicate with the outside world. A lot of data transmissions from the airplane. For them to all stop at the same time would take the work of somebody who has actually studied the systems in some detail to know how to turn off all of the systems at the same time."[300]

Peter Goelz, a former managing director of the US National Transportation Safety Board, said turning off a transponder is a "deliberate process. If someone did that in the cockpit," he said, "they were doing it to disguise the route of the plane. Was someone unauthorized inside the cockpit, ordering the transponders to be turned off and the plane to be turned around? Or did one of the pilots do it themselves?"[301]

Because it was considered highly plausible that Flight 370 had been hijacked, intense interest was immediately focused on its passengers.

"There was no distress signal or radio contact indicating a problem and, in the absence of any wreckage or flight data, police have been left trawling through passenger and crew lists for potential leads," reads an article in the *National Post*.[302]

"'Maybe somebody on the flight has bought a huge sum of insurance, who wants family to gain from it or somebody who has owed somebody so much money, you know, we are looking at all possibilities,' Malaysian police chief Khalid Abu Bakar told a news conference. 'We are looking very closely at the video footage taken at the KLIA [Kuala Lumpur International Airport], we are studying the behavioral pattern of all the passengers."[303]

Malaysia Airlines released the following breakdown on passenger information under its *Manifest for Flight 370*: "153 Chinese, thirty-eight Malaysians, seven Indonesians, six Australians, five Indians, four French, three Americans, two New Zealanders, two Canadians, two from Ukraine, one from Russia, one from Taiwan, one from Netherlands, and two Iranians who were found to be traveling under false documentation—one with an Italian passport and one with an Austrian passport."[304]

The two Iranian passengers with stolen passports were the primary focus of attention by investigators. "Both of the men using stolen passports were Iranian nationals: Pouria Nour Mohammad Mehrdad, 18, and Delavar Seyed Mohammad

Reza, 29."[305] Police and INTERPOL authorities were able to track the data. According to an ABC article:

"They say both men traveling with stolen passports arrived in Malaysia on February 28. The stolen passports had belonged to Austrian Christian Kozel and Italian Luigi Maraldi, neither of whom was on the plane. Both men have said their passports were stolen in Thailand—Kozel's in 2012 and Maraldi's in 2013. The BBC has reported that the men using the false passports purchased tickets together and were due to fly on to Europe from Beijing, meaning they did not have to apply for a Chinese visa and undergo further checks. An employee at a travel agency in Pattaya, Thailand, told *Reuters* the two had purchased the tickets there."[306]

Ronald Noble, then Secretary-General of INTERPOL, announced that the two passengers had traveled to Malaysia on their Iranian passports, and then apparently switched to the stolen Austrian and Italian passports.[307]

So, "How difficult is it to board a plane with a stolen passport?" asks a CNN article. "Not as hard as you might think." The article features an interview with the director of terrorism, intelligence, and security studies at Embry-Riddle Aeronautical University, Richard Bloom. "'On any given day,' says Bloom, 'many people travel using stolen or fake passports for reasons that have nothing to do with terrorism. ... They might be trying to immigrate illegally to another country or they might be smuggling stolen goods, people, drugs, or weapons, or trying to import otherwise legal goods without paying taxes,' said Bloom."[308]

However, the matter of the two Iranians traveling with forged documentation was dismissed by investigators, insofar as being linked to the flight's disappearance.

"The fact that at least two passengers on board had used stolen passports, confirmed by INTERPOL, has raised suspicions of foul play" reads the *National Post* article. "But Southeast Asia is known as a hub for false documents that are also used by smugglers, illegal migrants, and asylum seekers. Police chief Khalid said one of the men had been identified as a nineteen-year-old Iranian, Pouria Nour Mohammad Mehrdad, who appeared to be an illegal immigrant. 'We believe he is not likely to be a member of any terrorist group, and we believe he was trying to migrate to Germany,' Khalid said of the teenager. 'His mother was waiting for him in Frankfurt and had been in contact with authorities', he said."[309]

INTERPOL concurred with the assessment of Malaysian police, eventually determining that the Iranian passenger was trying to migrate to Germany and was not a potential terrorist. The conclusion reached by its then secretary-general, Ronald Noble, was that the disappearance of Malaysia Airlines Flight 370 did not appear to be related to terrorism. Noble said, "The more information we get, the more we're inclined to conclude that it was not a terrorist incident." A CNN article that considered the issue quoted this from Noble and paraphrased the consensus of the Malaysian authorities, saying: "There's no evidence to suggest either of the men traveling on the missing Malaysia Airlines flight was connected to any terrorist organizations."[310]

Security checks on the backgrounds of the other passengers cleared them as suspects, also, because there were no red flags, no links to terrorist activities. However, more extensive research revealed that twenty of the passengers on Flight 370 were employees of a high-tech firm, Freestyle Semiconductor.

Information related to that company's work has developed into a theory that will be explored later on: sabotage.[311]

All things considered, as author John Choisser poignantly observed, a scenario involving a hijacking or a terrorist "obviously requires a hijacker or terrorist, neither of which can be found in the manifest. So far a motive is missing also for that scenario."[312]

On April 2, 2014, Malaysian police announced that they had narrowed down the possibilities and <u>eliminated all passengers as suspects</u>, narrowing their inquiry to cabin and cockpit crew only.[313]

Also, it became evident that, Islamics with box cutters aside—yes, we are being sarcastic here, but only to those who actually believe the absolutely unrealistic official version of the 9/11 hijackings—there are much more modern ways to take control of an airplane these days. A discussion on hijacking would not be complete without exploring the world of advanced avionics, in which remote control of an aircraft is now a proven capability. We'll cover that subject—known as cyber jacking—later on.

Twelve

Explanation #3—Pilot Terrorist Activity, or Suicide

Malaysian authorities assertively, and very quickly, said they believed there was a good chance that Flight 370 had been "sabotaged on board," and that the strange things that had happened to it were "consistent with deliberate action by someone on the plane."[314] So, after they eliminated all passengers as potential hijackers, their focus shifted to the flight crew.

Since transponders on today's large commercial air-crafts don't turn off by themselves, and they almost never fail mechanically, it was widely assumed that human intervention was responsible for whatever happened to the flight. And this, in combination with the dismissal of any of the passengers as a terrorist, brought sharp focus on the flight crew. They were the ones in the cockpit and the ones at the controls. So, to many, it seemed obvious that they should be closely scrutinized. A *New York Times* article from March 17, 2014, reads:

> It is still a mystery what happened to the jetliner and the 239 people on board. But increasingly, the evidence shows that the disappearance was most likely not an accident, and that whoever changed the plane's flight path and turned off its communications systems had expert knowledge of the aircraft. And that has inevitably led investigators to look into the lives and backgrounds of the pilots.[315]

PILOT AS TERRORIST

Then, in June 2014, Captain Zaharie Shah was, formally and quite publicly, made the "prime suspect" by Malaysian police:

"After conducting 170 interviews, investigators noted strange behavior by the pilot. He had made no future plans— socially or professionally—and his home flight simulator was programmed with a flight path into the depths of the Southern Ocean before landing the plane on an island with a small runway. The drills were deleted from the computer but specialists were able to retrieve the files."[316]

That sounds pretty darn serious, doesn't it? "Prime suspect." They sounded pretty confident that they had a case against the pilot.

As noted in an article on the website Flight Club, Malaysian investigators "disclosed that files recovered from Captain

Zaharie Shah's flight simulator show its most recent path was set toward a small island in the southern Indian Ocean."[317]

That "small island" is also the home of a major US military base, Diego Garcia:

A March 19, 2014 article in the *Daily Mail* reads, "Today it was also revealed that a remote island in the middle of the Indian Ocean with a runway long enough to land a Boeing 777 was programmed into the home flight simulator of the pilot of the missing Malaysia Airlines plane. Police are now urgently investigating whether Captain Zaharie Ahmad Shah had practiced landing at Diego Garcia, an island south of the Maldives occupied by the US navy."[318]

People started "putting things together" very quickly—the connections were there to be made—or so they thought. Those final words from the cockpit—as originally reported—suddenly took on a haunting, ominous tone: "All right, good night."[319]

(It was later revealed that it was actually the copilot who had spoken those words.[320] Then the words were looked at in their fuller context, and they didn't seem ominous in the least bit. They seemed very normal, even mundane—standard operating procedure. It was merely a handing-off procedure, saying goodbye to Kuala Lumpur air traffic control prior to being picked up by Vietnam air traffic control.)

There was also speculation that "his home life was fraught with difficulties," though this has been denied by his family.[321] His "difficulties" and "no future plans" only made him a human being residing on planet earth, in the minds of most—but it was hard not to take note of the flight simulator's deleted files. One had to wonder why he would be practicing landing at a US military base.

Captain Shah's politics were also closely examined and were initially considered suspect.

He was related to Malaysian opposition leader Anwar Ibrahim, and had attended a court hearing just hours before the flight, in which Ibrahim was sentenced to prison.[322] So, some speculated that his politics had led him to some act of desperation. The fact that Captain Shah was pictured wearing a T-shirt with the slogan "Democracy is Dead" was hailed as evidence of his extremism.[323]

One assessment of the situation in June 2014 reads, "There are also claims that the 53-year-old was an extreme supporter of Malaysian Leader of Opposition Anwar Ibrahim, who was found guilty of sodomy in an appeals court just hours before Flight 370 went missing. Shah attended Ibrahim's trial, then immediately went to the airport. In addition, one of Shah's friends, also a pilot, told reporters in March that the captain had been under mental duress."[324]

However, while authorities continued to search with great seriousness for evidence of his "psychological instability," those who actually knew him painted quite a different picture of the man "renowned for his professionalism," who was a loyal husband and father of three children: "With his character, and what we knew of him, it just wouldn't make any sense that he would have anything to do with any sort of a deliberate action on his part," a friend said.[325] Other sane minds also eventually chimed in, and once those events were put into context—as is often the case—the official implications started to appear more and more confounded. It was learned that the "Democracy is Dead" shirt that, to some, had seemed so terrorism related, was actually in reference to a widely shared belief in Malaysia that an election had been "stolen." Choisser wrote:

"On the 8th of May, over 120,000 Malaysians rallied in protest over what they considered a 'rigged' election, which they won by popular vote, but was overruled by the ruling party on a technicality … They also object to a law which they say enables government persecution of political enemies. Because the government is based on Islamic law, homosexuality is a crime, and protesters claim that it is a convenient way to silence political foes. All protests and activities by this group have been peaceful."[326]

That was the real explanation for the T-shirt.

Independent investigators have also pointed out that the politics of Captain Shah were shared with *millions* of other Malaysians who, like Shah, also believed that an election had been stolen.[327] Therefore, the actions of Captain Shah were deemed to have been based on political interest, and nothing extreme. All Captain Shah was really guilty of, said friends, family, and coworkers, was a love of flying so pervasive that he made *YouTube* videos about his home flight simulator so that he could show it to the world. He was an "aviation fanatic" and he also cared about corrupt politics in his country.

Authorities then became very interested in the alleged (not that anybody actually *used* that word, but that's what it was, *alleged*) "roguish nature of the first officer," Fariq Abdul Hamid.[328]

A *National Post* article reads, "Malaysia Airlines … said it is investigating an Australian television report that the copilot on the missing plane had invited two women into the cockpit during a flight two years ago. Jonti Roos described the encounter on Australia's *A Current Affair* … Roos said she and her friend were allowed to stay in the cockpit during the entire one-hour flight on Dec. 14, 2011, from Phuket, Thailand, to Kuala Lumpur. She said the arrangement did not seem unusual to the plane's crew.

'Throughout the entire flight, they were talking to us and they were actually smoking throughout the flight,' Roos said ... The airline said it wouldn't comment until its investigation is complete."[329]

When the police finished with all their extensive background checks, what they had actually uncovered was very *un*suspicious information regarding both officers of Flight 370's flight crew. Captain Zaharie was a veteran pilot with more than 18,000 flying hours and had "a reputation as a conscientious and professional flyer."[330] Fariq was a young, congenial copilot and was known to be "on the verge of proposing to his girlfriend—who was also a pilot."[331]

"Fariq Abdul Hamid, 27, is seen as a less likely conspirator. He was due to marry his long-term girlfriend—a fellow pilot, from Air Asia—and said to be a mild-mannered man who occasionally attended his local mosque, and was a keen car enthusiast."[332]

The authorities held a magnifying glass to every single part of their lives, looking for any indication that there was a dangerous component somewhere in their personalities. But they didn't seem to have any. You can see the footage of the pilots going through security at Kuala Lumpur International, just prior to Flight 370's departure, on YouTube. They appear perfectly normal. Search online for "MH370 pilots going through security," or go to www.youtube.com/watch?v=1w2Esr6O_DE.

PILOT SUICIDE

By late June 2014, it was announced that police had cleared everyone on Flight 370, except the pilot.[333] According to an ABC News article:

"Allegedly, Shah's marriage was falling apart, and he was also having problems with his girlfriend. Shah's brother-in-law, Asuad Khan, denied those claims ... saying that Shah 'had a good life,' and the reports that implicate him in the plane's disappearance are false. 'I can see that a lot of people are saying a lot of things about him which is untrue. He had a lot of money, and he loved his daughter very much."[334]

One book about Flight 370, *Good Night Malaysian 370* by Ewan Wilson and Geoff Taylor, concluded that Captain Shah was "mentally ill" and that he committed suicide in an intentional act.[335] In the book, they "suggest pilot Zaharie Ahmad Shah committed murder-suicide by deliberately depressurizing the cabin and locking his copilot out of the cockpit, giving passengers only twenty minutes' oxygen supply."[336] The book cites several pilots who committed suicide and who they believe are good comparisons to Shah. Via the process of elimination, the book concludes that pilot suicide is the only serious option for an explanation.[337]

"Our book looks dispassionately and in depth at every possible alternative for what could have happened to MH370 on March 8," Taylor said. "We analyzed the possibilities of slow depressurization and hijacking and found that these were extremely unlikely," According to the *New Zealand Herald*, Taylor "said it was 'impossible' that the plane was shot down or suffered fire, electrical or catastrophic structural failure, or rapid depressurization ... [He] said several issues reinforced the book's claims, including the route MH370 took before disappearing, with its eight deliberate changes of course. He said murder-suicide was tragic and unthinkable, but 'sadly we think that is exactly what happened.'"[338]

However, Simon Gunson, whom *we* consulted on the matter, and who has studied the facts of Flight 370, strongly disagrees with that assessment and with the conclusions that those authors make. Gunson wrote:

"If the aircraft had a cabin fire at 35,000 ft. and the fire melted through the aircraft's hull then it is entirely feasible BECAUSE: decompression would snuff out a fire as quickly as it began, leaving the aircraft intact to continue flying. The book's author is not experienced on a Boeing 777 which is designed to autonomously restore electrical systems after power surges without human intervention."[339]

It should be kept in mind that some of those pilot "suicides" have been hotly disputed as such. Many experts disagree with those theories and have stated or implied that we should consider the possibility that sometimes a pilot suicide is simply a "convenient conclusion."[340] When the 1997 SilkAir Flight 185 that was en route to Singapore plunged directly into a river with the cockpit voice recorder turned off, American investigators deemed the crash deliberate, but Indonesian officials disagreed.[341]

In the case of Flight 370, common sense would seem to indicate that if the pilot had been intent on committing suicide, he would not have passed up the various opportunities he had to crash the plane. It's hard to wrap one's mind around why, if suicide was his intention, he allowed the plane to go far off course and fly like that for hours.

According to an article in the *New York Times*, "Some investigators are convinced that one of the pilots was involved, saying that no credible evidence has appeared for another explanation. But other say that the evidence suggesting pilot involvement is inconclusive and contradictory."[342]

Another article reads: "An FBI analysis of a homemade flight simulator taken from the home of the captain, Zaharie Ahmad Shah, found that many simulations had been deleted from it, including some to the Indian Ocean, but investigators said that was hard to interpret as evidence that Zaharie might have plotted a suicidal course. A clinical psychologist advising the inquiry has also been skeptical that anyone would commit suicide by flying for more than six hours with a planeload of people; previous cases linked to pilot suicide have involved pilots who appeared to fly almost straight down soon after takeoff."[343]

Retired US Air Force Lieutenant General, Thomas McInerney, echoed the exact same point, saying, "You're not going to fly into the Indian Ocean to crash it. You'd have crashed it in the Gulf of Thailand if [you] … wanted to do it."[344]

Aviation safety expert, John Cox, agreed. He said that "someone taking such pains to divert the plane does not fit the pattern of past cases when pilots intentionally crashed and killed everyone on board."[345]

All told, official investigations have indicated nothing suicidal or deranged about either member of the flight crew. They were first-rate pilots and exhibited no warning signings of suicide or serious character flaws. Choisser concluded from his independent investigation—which included a careful review of the pilots' actions in this case—that, "at worst, they appear to have acted adequately and professionally, under extremely adverse circumstances; and, at best, they were *heroic*." He concluded, "It would seem that pilot suicide is possible but highly unlikely, given the evidence at hand."[346]

Thirteen

Explanation #4—Mechanical Malfunction/Fire

One explanation that became popular in the weeks of uncertainty after Flight 370 was lost, was that an on-board electrical fire was responsible for its demise. A theory by Chris Goodfellow, an experienced pilot, that was widely disseminated, published first on Google+ and then by publications like *Wired*[347] and *The Atlantic*,[348] suggests that a fire may have been started by something as simple as an overheated tire on the front landing gear that started burning and smoking.

Here's how it went, according to *Business Insider*: "The pilots turned the plane toward an airport that could handle the 777, turned off the transponder along with other electronics in an effort to isolate the source of the fire, and were then overcome by smoke ... The plane's autopilot kept the course until it ran out of fuel and crashed hours later."[349]

That seemed like the simplest explanation which, many claim, is usually the one that is the most likely to have really happened. It may have sounded good to folks who read it on the Internet, but when they ran it all by other pilots, ones extremely familiar with the Boeing 777, they just weren't buying it.

In an interview with *Business Insider*, Michael G. Fortune, a retired pilot who now works as an aviation consultant and expert witness, said pilots preparing to change destination "would have communicated their emergency and intentions to turn around, as well as ask for assistance and direct routing to a suitable airport from the air traffic controllers very quickly."[350] But, again, the pilots didn't ask ATC for help. If there was a fire, wouldn't they have?

The theory in Goodfellow's article also states that the reason the transponder stopped working was because "in the case of fire the first response [is] to pull all the main busses and restore circuits one by one until you have isolated the bad one."[351] That could have knocked out the plane's transponder.

Fortune also disagreed with that. The *Business Insider* article reads, "Fortune, who flew 777-200ERs like the one involved here, said pilots follow a specific procedure when there's smoke or fire in the cabin. He didn't buy into the idea that the transponders would have been turned off in an attempt to deal with the problem. 'The checklist I utilized for smoke and fumes in the B-777-200ER does not specifically address the transponder being turned off,' he said."[352]

Another pilot they spoke with, Steve Abdu, a 777 captain for a major carrier, "echoed Fortune's point that there's a clear checklist to follow in this kind of situation. And, he pointed out, it's unlikely smoke would have knocked the pilots unconscious or killed them—because they have oxygen masks. Each pilot has a quick-donning mask, and putting it on is step one on the fire checklist. It covers the full face, even if the pilot wears glasses, and can be put on in about two seconds. "'These masks are quite excellent at protecting a pilot from smoke and fumes,' Fortune said. The masks are impressive. Pressing clips on the face part inflates the harness; letting go deflates it.'"[353]

As we covered in Chapter One, that special mask even contains a microphone for communication with air traffic control, communication absent in the case of Flight 370. If you'd like to better understand it, there's a video on YouTube that shows how sophisticated that pilot oxygen mask is, www.youtube.com/watch?v=f8aDRgY6D34#t=67.

Others questioned the logic of the fire explanation as well because it doesn't address that lingering question, Why would the transponder have been turned off? Many, resultantly, have rejected that idea and returned to the realm of some type of hijacking as their hypothesis, particularly after it was proclaimed that whatever happened to the flight was "consistent with deliberate action by someone on the plane" and that it continued flying for almost seven hours.[354]

An article in *Slate* reads, "Goodfellow's account is emotionally compelling, and it is based on some of the most important facts that have been established so far. And it is simple—to a fault. Take other major findings of the investigation into account, and Goodfellow's theory falls apart. For one thing, while it's true that MH370 did turn toward Langkawi and wound up over flying it, whoever was at the controls continued to maneuver after that point as well, turning sharply right at VAMPI waypoint, then left again at GIVAL. Such vigorous navigating would have been impossible for unconscious men."[355]

It's a valid, important point because the evidence *does indicate that the plane was being flown.* We know "the plane's systems were gradually switched off."[356]

The *Slate* article goes on to say, "Goodfellow's theory fails further when one remembers the electronic ping detected by the Inmarsat satellite at 8:11 on the morning of March 8. According to analysis provided by the Malaysian and United States governments, the pings narrowed the location of MH370 at that moment to one of two arcs, one in Central Asia and the other in the southern Indian Ocean. As MH370 flew from its original course toward Langkawi, it was headed toward neither. Without human intervention—which would go against Goodfellow's theory—it simply could not have reached the position we know it attained at 8:11 A.M."[357]

In fact, even Goodfellow, the person with whom that theory originated, reportedly became less certain about it over time; "Chris Goodfellow … revised his theory after the Malaysian government had presented the statement supporting the hijacking theory … and acknowledges he could be wrong."[358]

But here's the kicker—in the comments of his post, he admits that after Sunday's revelation of the hijacking theory from the Malaysia government, the fire theory may be wrong, saying: "I wrote this post before the information regarding the engines continuing to run for approximately six hours and the fact it seems ACARS was shut down before the transponder."[359]

As we noted in the previous chapter, the authors of at least one book have concluded that fire could not have been the problem that 370 faced. They maintain that, never in aviation history, have events similar to Flight 370 happened as a result of fire. In an exchange we had with Ewan Wilson, he maintained that he and Taylor researched all the available data:

"There were no cases where the fire started, then was extinguished or just stopped, and then the crew DID NOT then communicate and subsequently landed. The point is it would be unprecedented for a fire to start, take out the flight deck crew, take out the radio, ACARS, and transponder, and then for the fire to stop or be extinguished, and then for the aircraft to continue on flying for hours."[360]

However, Simon Gunson points out that fire as the result of a mechanical malfunction *could actually* be the culprit of Flight 370's problems. As Gunson points out, it's a whole different story with 370's aircraft because it was a Boeing 777, and "a Boeing 777 … is designed to autonomously restore electrical systems after power surges without human intervention. When the transponder blinked off," he wrote, "[it did] not mean the plane wasn't there, it [meant] … the transponder was no longer transmitting and that is consistent with an electrical fault."[361] If this still isn't clear, here's a more detailed explanation he gave us:

When modern airliners suffer electrical surges (and there are many examples I can cite to you) then they lose higher functions, like Nav/Comms [navigation and communication], but not more automatic functions. In this respect the electronic brain of a plane reacts much like a person does in a coma after a car crash; the self-aware thinking mind shuts down. But the heart does not lose rhythm and the body still performs lower functions. In an airliner, the transponder, ACARS flight management computer, etc., all hibernate in a comatose state. The B777 and B787 are uniquely designed to re-boot their electrical systems after a shut down. Older airliner designs do not have such a function. After Lumpur control were notified by Ho Chi Minh control at 1:38 A.M. MYT that Vietnam lost MH370 "off radar" at BITOD [waypoint], Malaysia Airlines informed HCM control they believed it was still flying over Cambodia. If the aircraft had a cabin fire at 35,000 ft. and the fire melted through the aircraft's hull, then it is entirely feasible BECAUSE: decompression would snuff out a fire as quickly as it began, leaving the aircraft intact to continue flying.[362]

Gunson also had a very noteworthy explanation on the issues of radar coverage and the lack of a distress call:

> There is very little radar coverage over Vietnam, unlike IGARI where the turn-back was claimed to happen. And, according to a report in the Taiwanese paper the *China Times,* 8th March, they did make a distress call, saying they wanted an immediate landing because their cabin was disintegrating.

We detail the above report of a distress call in the final chapter of this book.

Gunson also makes an important point in a message board on the *Guardian's* site concerning the possibility of a cargo fire, which has been cited as a strong possibility due to the fact that the cargo hold of Flight 370 was known to contain a large quantity of highly flammable batteries. He notes that the color of the fire seen by the most reliable witness, oil rig worker Mike McKay, was not the correct color for a Lithium-ion battery fire:

"I have an unpublished e-mail from the oil rig worker, Mike McKay, in which he described there was a bright orange flame visible down the left side of the plane. If the fire was caused by Lithium batteries in the cargo then the color of the flame would be violet, so this suggests there was no cargo fire."[363]

John Choisser also dismissed the likelihood of a cargo fire, writing:

"If there was a developing emergency, such as a cargo fire, the flight crew would have been alerted to take immediate action, and the problem reported before the airplane left Malaysian ATC space. The same could be said for an intruder, or for someone gaining access to the E/E bay through the floor panel."[364]

Note the paradox here: This conclusion seems to negate the possibility of a slowly developing emergency—like a fire or a hijacking—and logically implies that something catastrophic and immediate in nature had to have occurred to Flight 370. However, other evidence—such as the total absence of debris and a plane that continued flying for many hours—seem to negate the possibility of a catastrophic event, like massive mechanical failure, a bomb, or a missile shoot-down.

According to the practice of the philosophy of logic, when you run into a seeming paradox, you just have to follow the facts. You reexamine the facts, because something is "rotten in Denmark" somewhere; and you probably have an incorrect premise. So—we followed the evidence and, oddly enough, like the aviation consultants, it runs in opposite directions. On the one hand, fire seems very unlikely because of what did *not* happen. Typically, a fire is immediately reported. At the same time, fire is about the only thing that comes close to a simple explanation that fits the evidence. Aircrafts are so safe today that catastrophic mechanical failure has almost been completely engineered out of the equation. As Chris Lee, an aerospace engineer, observed in a message board on Quora:

"Aircraft[s] are generally so safe today that only a completely unforeseen failure or a total breakdown in the pilot-aircraft interface could plausibly cause an accident in high altitude cruise ... Catastrophic events occurring during the cruise phase of flight are exceedingly rare in today's world. Given the altitude, there are few opportunities for weather or piloting deficiencies, and even something like an engine failure will still provide you with many options."[365]

And, please note, that one of those "many options" cited above would certainly be declaring an emergency.

Sid McGuirk, an associate professor of air traffic management at Embry-Riddle Aeronautical University, had this to say following the disappearance of Flight 370:

> The flight crews use combinations of high-frequency radio, satellite-based voice communication, and text data networks to report to ATC (air traffic control) the exact time, position and flight level when the crossing begins.

The *21st Century Wire* sums it up well, saying, "Based on this information, it seems that the public has not been given the full story as to what has happened to the missing flight, making it unlikely that the event was 'sudden and catastrophic,' as the plane apparently turned around according to initial reports, traveling some 350 miles and making it to

Malacca Strait, according to Malaysian officials. It would have taken fifteen to thirty minutes for the plane to land according to reports, leaving a ghost pattern on radar as to their location and plenty of time for a distress signal if the plane wasn't hijacked."[366]

That is an extremely important point because:

"Air traffic controllers would have known the exact moment that something had changed during the course of this flight."[367]

So, the plane was apparently being flown. *Reuters* quoted a senior military officer, who was briefed on the investigation, in a piece they wrote about this, "It changed course after Kota Bharu and took a lower altitude. It made it into the Malacca Strait." That would appear to rule out sudden catastrophic mechanical failure because it would mean the plane flew around 500 km at least after its last contact with air traffic control, although its transponder and other tracking systems were off."[368]

Also, one more reason can be added as to why fire may seem an insufficient explanation: there would have been no reason to cover anything up, and it seems there was quite a bit of shenanigans in that respect. If it was actually fire or major mechanical malfunction, it could have simply been reported. You know, like—Telling The Truth—there's a novel concept, huh? Instead, it seems there has been an intentional wild goose chase and this global game of Marco Polo. The point is that the authorities could have immediately squelched all the rumblings about electronic hijacking and Diego Garcia possibilities with just a simple admission that it looked like fire was to blame. Instead, the political gamesmanship, on a global basis, has truly clouded the issue, to the point where no one is even sure what ocean they should be looking for it in.

MECHANICAL FAILURE

Mechanical failure was, according to a *Telegraph* article, "understandably, the initial assumption." But, the article goes on to say, "America quickly said that its satellites had detected no explosion, and neither Rolls-Royce—the engine manufacturers—nor Boeing, the aircraft designers, seemed to think there had been a mechanical failure."[369]

And Malaysian authorities maintained that there was no further communications from the crew. This seems pretty odd if you ask us, because even with an on-board fire, why wouldn't the pilots or anyone on board be "attempting to call home"?[370]

Here's one answer:

That assumption, that there was no distress call, well be **the incorrect premise** of this whole crazy equation. Recall that *The China Times* reported that there *was* a distress call. If that's true and, for whatever reason, it's being obfuscated by the authorities, that change would then make the equation a more logical one.

And then we found something big—something that has been under *everyone's* radar: Highly sensitive infrasonic atmospheric sensors can pick up sound waves from a wide variety of sources, including rocket launches and aircraft crashes.[371] Infrasonic events around the globe are constantly monitored by a watchdog agency of the United Nations, primarily to guard against nuclear and missile testing: "The Comprehensive Nuclear-Test-Ban Treaty Organization (CTBTO) says they can detect explosions and impacts of aircraft on water."[372] That relates importantly to Flight 370, in regard to what *they didn't* hear:

> **Spokesman for UN Secretary-General Ban Ki-moon, Stéphane Dujarric, told reporters in New York yesterday [March 17, which was <u>nine days after</u> Flight 370 disappeared]: [that] 'Regarding the missing Malaysian Airlines flight ... The Vienna-based Comprehensive Nuclear-Test-Ban Treaty Organization (CTBTO) confirmed that neither an explosion nor a plane crash on land or on water had been detected so far.[373]**

So that would clearly indicate that something *other than* what we have been told happened to Flight 370. That's a huge factor: If it crashed, as we have been told, then why wasn't it picked up by those sophisticated listening stations? Bear that fact in mind, especially as you consider the points in the upcoming chapter on advanced technologies.

Fourteen

Explanation #5—Aliens/Other Extraordinary Means

Before you scoff at this idea and immediately dismiss it as utter nonsense, it would probably be a good idea to read a recent statement by NASA (National Aeronautics Space Administration). They tell us it is a virtual certainty that not only will we <u>discover alien life out there, we will discover it within the next twenty years</u>.[374]

Here is what the head honcho at NASA had to say a short while ago:

> It's highly improbable in the limitless vastness of the universe that we humans stand alone. I think in the next twenty years we will find out we are not alone in the universe.[375]

Think about *that* one for a while, Charlie Brown. And I'll bet his telescope is better than yours!

Personally, we haven't had any direct experience with alien life forms (not counting strange friends and acquaintances), at least not in recent memory. Well, there was that time in the men's room at Yankee Stadium back in the seventies ... but *everybody* was experimenting with drugs back then!

General Nathan Farragut Twining was a big shot in the United States Air Force during the 1940s and 1950s. After being promoted from Lieutenant General (three stars) to full General (four stars) and Vice Chief of Staff of the Air Force,[376] in 1947 he was assigned the task of assessing the credibility of numerous reports of flying saucers. To place that in historical context, it was two years after the Roswell Incident (we'll *get* to that in a minute) and other UFO incidents near the same time, in which there appeared to be evidence of alien flight and even tangible debris left.[377] General Twining fully investigated these claims and wrote an astounding memorandum (we'll get to *that* also) on his recommendations concerning the reports of "Flying Discs."

Before you read the general's formal memorandum—lest you think he was dismissed as a loony or a charlatan—consider the following: At the time of his memorandum—dated September 23, 1947—General Twining was Chief of the US AMC (Air Materiel Command) and, as Vice Chief of Staff, the second most powerful officer of the US Air Force. *After* his memo, his recommendations were followed and took form in the special programs *Project Sign*, which became *Project Grudge*, and then the infamous *Project Bluebook*. General Twining was fast-tracked to Chief of Staff of the United States Air Force in 1953 (as soon as that post was available) and then appointed by

President Eisenhower to Chairman of the Joint Chiefs of Staff in 1957—the most powerful post in the entire United States Armed Forces and, in fact, the senior military advisor to the President of the United States.[378] Not exactly a kook! Below are the portions of that memo in the National Archives—marked "SECRET"—that struck many as being extraordinary.

1. As requested by AC/AS-2 there is **presented below the considered opinion of this command concerning the so-called 'Flying Discs.'**...
2. It is the opinion that:
    a. **The phenomenon is something real and not visionary or fictitious.**
    b. There are objects probably approximating the shape of a disc, of such appreciable size as to appear to be as large as man-made aircraft.
    c. There is a possibility that some of the incidents may be caused by natural phenomena, such as meteors.
    d. The reported operating characteristics such as extreme rates of climb, maneuverability (particularly in roll), and motion which must be considered <u>evasive</u> when sighted or contacted by friendly aircraft and radar, lend belief to the possibility that some of the objects are controlled either manually, automatically, or remotely.
    e. The apparent common description is as follows:
        (1) Metallic or light reflecting surface.
        (2) Absence of trail, except in a few instances when the object apparently was operating under high performance conditions.

(3) Circular or elliptical in shape, flat on bottom and domed on top.

(4) Several reports of well kept formation flights varying from three to nine objects.

(5) Normally no associated sound, except in three instances a substantial rumbling roar was noted.

(6) Level flight speeds normally above 300 knots are estimated …

It is recommended that:-

a. **Headquarters, Army Air Forces issue a directive assigning a priority, security classification and code name for a detailed study of this matter** to include the preparation of complete sets of all available and pertinent data which will then be made available to the Army, Navy, Atomic Energy Commission, JRDB, the Air Force Scientific Advisory Group, NACA, and the RAND and NEPA projects for comments and recommendations, with a preliminary report to be forwarded within 15 days of receipt of the data and a detailed report thereafter every 30 days as the investigation develops.[379]

From this, Project Blue Book was born. The "Roswell Incident" apparently had credibility, and at the highest levels of the US government:

"*July 2, 1947:* On this date a flying saucer abruptly stopped flying over Roswell, New Mexico. The wreck and the subsequent shoddy cover-up is believed to have launched a million government skeptics, countless researchers, and innumerable theorists, some of whom believe that at least one survivor and the bodies of several other extraterrestrials were recovered from the crash."[380]

Other incidents near that event—geographically and chronologically—also attracted some serious attention. Here are some of the more well-known:

February 13, 1948: When three radar units tracked a nose-diving aircraft northeast of Aztec, New Mexico, Secretary of State George Marshall asked that a party of investigators search the area. They discovered a demolished thirty-foot disc and, some believe, the remains of a dozen charred humanoids.

May 21, 1953: An Air Force helicopter pilot reports that he was summoned to the scene of a UFO crash in Kingman, Arizona. The pilot, who had been recruited to help in the recovery of the craft, claimed to have seen a large, metal disc that had impacted the ground so hard it plowed up a furrow of earth some twenty inches deep. In an affidavit published in *UFO* magazine, he also claimed that a space-suited alien was found in the wreckage, dead.[381]

General Twining also flatly denied that the "flying disks" were the product of any US military experimental projects of any kind.[382]

Project Blue Book was not actually blue, or even a book. It was a compilation, done in secret by the US military—said to be 140,000 pages long—of "information relating to various UFO sightings that were called or mailed in to Project Blue Book headquarters."[383] Plus, of course, whatever they came up with on their own. It all sounds like an episode of the *X-Files*, but it's not anything we're going to actually hear about, folks.

It would be interesting to know what they concluded from all that. However, in typical government fashion, they just buried it all. Even senior United States Senator Barry Goldwater couldn't get these military types to release their findings. Stonewalled by the military in his attempts to gain access to UFO information that Senator Goldwater believed was housed at Wright-Patterson Air Force Base, he wrote the following:

> The subject of UFOs is one that has interested me for some long time. About ten to twelve years ago I made an effort to find out what was in the building at Wright-Patterson Air Force Base where the information is stored that has been collected by the Air Force, and I was understandably denied the request. It is still classified above Top Secret. I have, however, heard that there is a plan underway to release some, if not all, of this material in the near future. I'm just as anxious to see this material as you are, and I hope we will not have to wait much longer.[384]

The information was not released—not to Senator Goldwater or, as far as we know, to anyone else.

Senator Goldwater and another senior Senator, Richard Russell, were reportedly fervent believers in UFOs. Senator Russell reportedly even *saw* one.[385] Senator Goldwater reportedly told former astronaut Clark McClelland that "the UFO situation is the highest level of national secrecy."[386] Several experienced US astronauts are in agreement that UFOs are real.[387] They are trained observers and they stand by what they have seen with their own eyes. You can read about their observations here, www.stargate-chronicles.com/site/quotes-by-those-who-know/.

A lot of other very high-profile people are also believers—in fact, the list will surprise you:

President Richard Nixon, President Ronald Reagan, President Jimmy Carter, Soviet Premier Mikhail Gorbachev, FBI Director J. Edgar Hoover, General Douglas MacArthur, and Professor Stephen Hawking.[388]

The government of Chile apparently agrees with them. A Chilean study was released to the public in 2014. It was a highly scientific analysis of photographs and sightings at a remote copper mine in the Andean plateau, fourteen thousand feet above sea level.

The study concludes that "It is an object or phenomenon of great interest, and it can be qualified as a UFO."[389] The photo analysis can be viewed at www.openminds.tv/chilean-government-cant-explain-ufo-photo/28846.

Another sighting in Chile, in the same year, took place over a large water reservoir in a remote area of the southern part of the country. The director general of the organization that investigates such matters in Chile, a retired military general, announced that the conclusions of their study were not only that it was an authentic UFO, but estimated its size as equal to two soccer stadiums. He said:

> After considerable research time we reached a series of conclusions that are the same obtained in the United States, and which are: this photograph is real and not a hoax; Second, that the incidence of the light in these clouds is the same as that which falls upon the object; Third, that it has its own light. And therefore a series of portholes are visible. This is according to our Ph.D in meteorology, and according to the clouds existing at the time during

that season in the Cordillera, makes it twice the size of the National Stadium (in Chile).

We do not know what it is or where it came from, but the anomalous aerial phenomenon described as an unidentified flying object is real and we have the proof and the eyewitness accounts to support it,' notes [General] Bermudez, who was a military pilot for the Chilean Air Force, flying F-5 fighters among other craft.[390]

Former astronaut Clark McClelland was a certified SpaceCraft Operator and a director of NICAP (National Investigations Committee on Aerial Phenomena), which has been referred to as "the real *X-Files*." He has testified that he actually personally witnessed an alien being who was attempting to communicate with the Space Shuttle crew. This is his affidavit, reproduced word for word:

I, Clark C. McClelland, former ScO, (Spacecraft Operator) Space Shuttle Fleet, personally observed an 8 to 9-foot tall ET on a LCD 27-inch video monitors while on duty in the Kennedy Space Center, Launch Control Center (LCC). The ET was standing upright in the Space Shuttle Payload Bay having a discussion with two (2) tethered US NASA Astronauts!

I also observed on my 27-inch monitors, the spacecraft of the ET as it was in a stabilized, safe orbit in the background, to the rear of the Space Shuttle main engine pods. **I observed this incident for about one minute and seven seconds. That is plenty of time to memorize all that I was observing: An ET and Alien Star Ship.**

A friend of mine later contacted me and said that he had received an email from a person who also had observed an 8 to 9 foot tall ET inside the Space Shuttle Mid Deck Crew Compartment! Yes, inside our Shuttle! **Both these missions were Department of Defense (DoD) Pentagon Top Secret encounters!**

My background is easily verifiable. There is no Federal Government Agency that can dispute my credentials, my experience or my presence at Kennedy Space Center at that time! I am a Space Program Pioneer. I assisted in launching the Mercury, Gemini, Apollo, Apollo-Soyuz, Skylab, Space Shuttle, Deep Space Missions, and the International Space Station. I have launched or witnessed 650 missions, so far in my life!

I have received testimonials to my career and character from: Walter Cronkite, Major Donald Keyhoe, NICAP Director, Richard Hall, Assistant Director of NICAP, Astronomer/UFO & past US Government Agent, Dr. J. Allen Hynek, LCC launch site team mate, etc.

I served as the Assistant State of Florida and KSC Director of MUFON, and the Director of the NICAP (National Investigations Committee on Aerial Phenomena) at Cape Canaveral and KSC!

I have received honored mentions from US Senators, Congressmen, Military Officers and Scientists.

My testimony supports what many astronauts have seen and some have reported. [Watch the NASA Live Transmission video link below.]

I am an expert in visually recognizing any crafts created and flown by the human race, whether secret or otherwise! I know an ET and Alien craft when I see them! I was the Director of the NICAP Unit-3 actual X-Files at Cape Canaveral and the Kennedy Space Center, 1958 to 1992! Yes, the actual X-FILES!! Aliens are here!! They walk among US! They may have infiltrated our various earth governments!!

What I know is that "they" (NASA) would not allow me to publish my story for many years! I have written at least three books to release the truth to humanity. I am on the leading edge of research in this subject, and I have met with many NASA people and astronauts throughout the years that I have been involved in Military/NASA Space Programs. I have heard of many experiences of ETs that were seen in space, on the Moon, from some astronauts and I relate these accounts and more in my books.

NASA is not a civilian space agency! The Pentagon owns NASA. Some of the DoD [Department of Defense] missions I participated in were classified 'Top Secret' (TS).

Those missions carried TS Satellites and other space mission hardware into orbit where several crews met with ETs!

I am ready to tell my story.

Clark C. McClelland,

Former ScO (Space Craft Operator)

Space Shuttle Fleet, Kennedy Space Center, Florida.[391]

The above is *not* a hoax, in case you are wondering. McClelland is very professional and has a very informative website. Check it out for yourself, www.stargate-chronicles.com/site/.

Well, what do you think of them apples? Maybe you should read those last few paragraphs again, huh? *We did.* That would certainly explain a lot—like that *unearthly* look in the eyes of a lot of our politicians! Actually, have you ever noticed that Condoleezza Rice and Dick Cheney always—well, never mind.

McClelland also approached the topic with former Senator Barry Goldwater at Kennedy Space Center in 1969:

> I seriously pondered my next question concerning UFOs and finally asked him if he would discuss the subject with me? I was surprised he had no Secret Service Agents or NASA Public Affairs personnel hovering around him. I explained my position at KSC [Kennedy Space Center] and my being the NICAP Unit-3 Director for Major Donald E. Keyhoe [International Director, NICAP] at KSC. He had met Keyhoe in DC (Washington) years before. He approved our discussion on the UFO subject and we began our exchange of information. I began by saying it was obvious he had a deep abiding interest in UFOs and possible visits of alien races to earth. He said, That is accepted fact in Washington and especially at the Pentagon, young man.[392]

So, how does the possibility of alien life relate to Malaysia Flight 370, one might ask? That's a *damn good* question. We have no idea. It's just that it's something that a discernible percentage of the population actually believes in and—if it is true—it's an unknown that some apparently believe could possibly explain many otherwise unexplainable things, including the bizarre circumstances of Flight 370.

Therefore, all things considered, it appears one has to conclude that aliens are a "possible"—it's just *so* hard to get an interview with one!

Fifteen

Explanation #6—Shoot-Down by Missile

Because Flight 370 reportedly disappeared off radar very quickly and there was apparently no time for a distress call, speculation quickly went to a possible missile shoot-down. The fact that "war games" were taking place at that time lent further credence to the idea. Two huge joint military exercises were taking place at the same time that Flight 370 vanished from radar:[393]

"What puzzled me most since the 'disappearance' of MH370," wrote Matthias Chang of the website *Future FastForward*, "is the deafening silence of the military establishments of the United States, Thailand, Singapore (and to a lesser extent those countries who are mere observers) who prior and subsequent to the 'disappearance' of MH370 were involved up to their eyeballs in the annual military exercises 'Cobra Gold' and 'Cope Tiger' led by the United States beginning from 11th February and ending 21st March, 2014.'[394]

The Cope Tiger 2014 Field Training Exercise was a joint effort of the United States Air Force, Singapore Air Force, and Thailand's Air Force, Army, and Navy.[395] Cobra Gold was a Field Training Exercise between the United States, Thailand, South Korea, Japan, Malaysia, and Singapore.[396] Atop that, since no mayday was declared—there was no declaration of an in-flight emergency of any kind by any member of the flight crew of 370—it seemed that a catastrophic incident must have knocked the plane out of the sky.

Fact: So far, no debris field of plane wreckage has been linked to the [Boeing] 777, which would indicate a bomb blast.

Analysis: When Robert Francis, former vice chairman of the US National Transportation Safety Board, heard about the missing plane, his immediate thought was: 'For some reason the aircraft blew up and there was no signal, there was nothing.' The fact that the plane disappeared from radar without warning indicated to Francis 'there was something unprecedented that hasn't happened before.'[397]

So, a missile strike, to many, seemed to make sense, given the circumstances initially known. In fact, veteran national security analyst Michael Shrimpton presented a very interesting scenario on how, and under what actual circumstances, Flight 370 could have also been shot down by a missile fired by the Chinese.[398] Check it out here: www.veteranstoday.com/2014/03/15/the-

shootdown-of-malaysian-airlines-flight-mh370/
comment-page-1/.

Author Nigel Cawthorne, in his book *Flight MH370: The Mystery*[399] claims that a missile strike matches the evidence, especially the testimony of the oil rig worker, Mike McKay, which we profiled earlier. An article in the *Mirror* reads:

> The book claims the drill was to involve mock warfare on land, in water, and in the air, and would include live-fire exercises. It adds: "Say a participant accidentally shot down Flight MH370. Such things do happen. No one wants another Lockerbie, so those involved would have every reason to keep quiet about it." But he goes on to support one theory, based on the eye-witness testimony of New Zealand oil rig worker, Mike McKay, that the plane was shot down shortly after it stopped communicating with air traffic controllers. At the time there was a series of war games taking place in the South China Sea involving Thailand, the US, and personnel from China, Japan, Indonesia, and others. New Zealand oil rig worker Mike McKay also claims to have seen a burning plane going down in the Gulf of Thailand.[400]

As time went by, however, it was pointed out that a missile would have left a "signature" that would have been tracked by radar stations. It was further observed that there would have been some amount of debris from the disintegration of the aircraft after a missile impact and that debris should have been clearly visible in the immediate search efforts, and, reportedly, at least, it was not.

Then, Flight 17 was shot down by a missile over Ukraine and there was a vivid example of a missile shoot-down. The evidence of Flight 17 did not compare to Flight 370, especially insofar as the radar signature of the missile, and a large amount of visible debris. So there was, therefore, much less of a proclivity toward the possibility that it was a missile shoot-down that brought down Flight 370, because "we saw what happens" during a missile shoot-down with the example of Flight 17 in Ukraine.

Then, in October 2014, the foreign minister of the Netherlands, Frans Timmermans, revealed that even though it had been widely reported that Flight 17 disintegrated on contact with the missile, one of the passengers on Flight 17 apparently had time to put on an oxygen mask. The minister was dramatic about the implications of that point:

"No, they did not see the missile coming, but you know that someone was found with an oxygen mask over his mouth? So he had time to do that?"[401]

As CNN observed, "That would imply the passengers weren't killed instantly when the plane was downed in July over Ukraine, reportedly by a missile fired by pro-Russian rebels."[402]

David Soucie, aviation expert and author of the book *Why Planes Crash*, also sounded somewhat surprised about the oxygen mask on a passenger:

> People think of a missile as one projectile. But when this missile explodes, it's before it hits the aircraft, that it explodes, and it sends thousands of small pieces of debris through the aircraft, just as though they were bullets going through the aircraft. So the fact that someone, or *any*one, had survived the initial impact, is pretty unlikely but, the fact is, it is possible.[403]

So, even the flight that was shot down by a missile—Malaysia Flight 17 over Ukraine—has some things that don't seem to add up. As far as Malaysia Flight 370 is concerned, the possibility that it was a missile strike is not currently the opinion of most experts.

Colonel J. Joseph is a former pilot and an aviation accident consultant at Joseph Aviation Consulting in Texas. Colonel Joseph discounted the possibility of Flight 370 having been brought down by a missile, for the following reasons:

> I would be very surprised at this point, however, if the intelligence communities throughout the world were not a little more aggressive in pointing something like that out. Most of the world is under this surveillance in some way, shape, form, or another, and certainly satellite imagery would be something that would pick up a bright flash like that had that occurred in the night sky over a dark ocean. So I think we would have seen some preliminary intelligence data at least pointing us in that direction, had that been the occurrence.[404]

On the *other* side of the coin, one could say that same "satellite imagery would be something that would pick up" something as huge as a Boeing 777, yet they supposedly lost all track of it, if the official reports are to be believed.

We talked to one of *our* analysts, among whom was veteran US Intelligence operative William Robert "Tosh" Plumlee, whom we have consulted numerous times for our books because he is extremely astute on international intelligence matters.

In a highly specific post on a popular social media site—weeks before the event occurred—Plumlee cited a NATO

INTEL briefing report obtained from two international intelligence sources whom he and others within the US Intelligence Community considered extremely credible, specifically predicting "that a 'shoot-down' test firing was in the works, soon to be tried on an airliner or cargo type aircraft in the UKRAINE. The testing of the system would be launched within this time frame, indicating a test firing of their new missile radar system and BUK missiles obtained from Russia." The following statement is from Plumlee. His statements were mentioned and posted on a social media network a few weeks before the event happened.[405] They read:

> Correct. I (Plumlee) did predict the shoot down of MH17, before the actual event. I even made mention of the type system that would be used. That does not mean I am clairvoyant. That means I had received prior information—before the event. This intel information which I had received, came from two very reliable sources who were working within a secret intelligence unit assigned to NATO operations within the Middle East.[406]

It may be surprising to many that the motives "in play" for all of that were actually the huge business of international gunrunning. As Plumlee also stated in a related post:

> This is all part of an international weapons market (Russia, USA, Saudi, and Qatar weapons agreements). *Other* U.S intelligence sources within the CIA/Pentagon had also received the same information prior to the shoot-down of Flight 17 over Ukraine, but were blocked by the Obama administration because of the mutual agreements between

Russia and the US; *thus the phone call* with Putin and Obama shortly after the shoot-down.

Yes, the NATO-leaked intel regarding the Ukraine shoot-down, and the Plumlee postings about two weeks before the event, were vetted and proved to be right on target.[407]

Those same intelligence sources indicate that Flight 370 may have also been a casualty of the international arms trade, as the result of an errant missile. As Plumlee put it:

The NATO intel sources are also checking the Malaysia 370 flight and they believe there was a testing failure that locked on to Flight 370 by accident, back in March. That too was reported via proper channels to various international sources but was also blocked, in order to protect the weapon shipments going to the Middle East.[408]

If that is true—that a missile brought down Flight 370, accidentally or otherwise—it would mean that all the silence, unnecessary delay, and misdirection regarding the search area, were actually part of a larger plan to discreetly recover and destroy what little evidence of the aircraft remained after the strike.

Sixteen

Explanation #7—Sabotage

One of the first theories that developed regarding the missing aircraft was also one of the most provocative. As one reporter observed, regarding the mystery of Flight 370:

"Two aspects not reported regarding the mystery are: 1) Using today's electronic weaponry, a plane can seemingly 'vanish,' and 2) Passengers aboard the missing Malaysia jet linking to contracts with the Department of Defense and high-tech electronic weaponry. Today's electronic warfare (EW) capability includes weaponry that can hide planes. Electronic weaponry is not only available, it is being deployed. Is this being used to hide or 'cloak' the 'vanished' plane?"[409]

It was learned that twenty of the passengers on the flight were employees of a high-tech company with links to the military. Here's what we know:

A US technology company which had twenty senior staff on board Malaysia Airlines Flight MH370 had just launched a new electronic warfare gadget for military radar systems in the days before the Boeing 777 went missing. Freescale Semiconductor, which makes powerful microchips for industries including defense, released the powerful new products to the American market on March 3. Five days later, Flight MH370 took off from Kuala Lumpur for Beijing with 239 people on board including twenty working for Freescale. Twelve were from Malaysia, while eight were Chinese nationals. Freescale's spokesman Mitch Haws has said, 'These were all people with a lot of experience and technical background and they were very important people.'[410]

And, with the twenty high-tech employees, *along* came a theory:

Experts have been baffled how a large passenger jet seems to have flown undetected and possibly beaten military radar systems for up to six hours. Avoiding radar via 'cloaking technology' has long been one of the objectives of the defense industry and Freescale has been actively developing chips for military radar. On its website, the company says its radio frequency products meet the requirements for applications in 'avionics, radar, communications, missile guidance, electronic warfare, and identification friend or foe.' Last June it announced it was creating a team of specialists dedicated to producing 'radio frequency power products' for the defense industry. And on March 3, it announced it was releasing eleven of these new gadgets for use in 'high frequency, VHF and low-band UHF radar and radio communications.'[411]

Freescale Semiconductor did not respond to the numerous reports on Internet forums and as comments to articles. But the buzz was big. Here was the origin:

Have you pieced together the puzzle of missing Flight 370 to Beijing China? If not, here are your missing pieces. Patents, Patents, Patents. Four days after the missing Flight MH370, a patent is approved by the patent office; four of the five patent holders are Chinese employees of Freescale Semiconductor of Austin, TX. Patent is divided up on 20% increments to five holders: Peidong Wang, Suzhou, China, (20%); Zhijun Chen, Suzhou, China, (20%); Zhihong Cheng, Suzhou, China, (20%); Li Ying, Suzhou, China, (20%); Freescale Semiconductor (20%). If a patent holder dies, then the remaining holders equally share the dividends of the deceased, if not disputed in a will. If four of the five dies, then the remaining one patent holder gets 100% of the wealth of the patent. That remaining live patent holder is Freescale Semiconductor. ... Here is your motive for the missing Beijing plane. As all four Chinese members of the patent were passengers on the missing plane. Patent holders can alter the proceeds legally by passing wealth to their heirs. However, they cannot do so until the patent is approved. So when the plane went missing, the patent had not been approved.[412]

The theory goes from there, making links that suggest that since Freescale is owned by The Blackstone Group, and the Rothschilds "control" Blackstone, the trail of blood goes right to the top of the food chain:

"In laymen's terms, Lord Jacob Rothschild is now the patent holder by virtue of invested interest into Freescale Semiconductor Ltd. To bring things further into perspective, putting the icing on the cake, the Rothschild dynasty owns the Malaysian Central Bank, which, in-turn, is heavily invested into the Malaysian government and Malaysia Airlines."[413]

Well, as you may have suspected, it's not quite that simple. For one thing, the names mentioned as the patent holders are not a match to the names on the passenger manifest for Flight 370:

"Although a Freescale patent does exist under number US8650327, none of the names listed actually appear on the passenger manifest released by the Malaysian authorities."[414]

Some of the names are very close, but not a direct match. Therefore, as one researcher observes, *it's possible* the absence of some names "could be due to slight differences in the transliterations of those persons' names from Chinese to English."[415]

But the theory does not at first seem to hold up to scrutiny.

"Moreover, the patent in question is dated 11 March, 2014, and involves a 'system for optimizing number of dies produced on a wafer.' That doesn't sound like a highly valuable, 'breakthrough' type of patent that would prompt the murder of four people (much less the death of 235 other innocent parties) in order to gain exclusive control of it."[416]

On top of that, the succession of the legal rights regarding the patents is somewhat suspect as well.[417] It does not seem to be as simple a matter as the theory implies—there are rights of inheritance and a million or so legal issues to address. *And*, as it's been further observed, the "Right Honorable Lord Rothschild"—which is actually his title, believe it or not—is merely an *advisor* to The Blackstone Group, as anyone can plainly see, right on their website, www.blackstone.com/the-firm/overview/our-people/jacob-rothschild.

On the *other* hand, it's not *quite* as crazy as it appears. We did some research and learned that Freestyle Semiconductor was indeed bought out by The Blackstone Group, for a whopping $17.6 billion dollars, back in September 2006.[418] The Carlyle Group was also a huge investor in Freescale, leading the buyout consortium that included the notorious Blackstone Group. The Carlyle Group may "ring a bell," because they have some very heavy hitters on the geopolitical field and have basically been called a revolving financial door for everyone from George Bush to bin Laden.[419] The Carlyle Group is also linked to Dick Cheney, Halliburton, and the "military-corporate complex" insofar as their alleged ability to make obscene profits from war.[420]

And, apparently, Lord Rothschild *really does* have some substantial business interests in Malaysia. It's reportedly true that the Rothschild family actually has a "significant investment" in Malaysia Airlines, via their reported control of the Malaysia Central Bank.[421] *And*, contrary to the report just cited, that microchip invented by Freescale *really does* appear to be a big deal in the world of military technology.[422]

In a detailed report from *Malaysia Chronicle* dated April 8, it said that Freescale launched what could be the world's smallest microcontroller in Feb 2013 called the Kinesis KL02.

KL02 measures 1.9 mm by 2 mm and contains RAM, ROM, and a clock. Even with its minute size, KL02 might be the most potent next-generation war weaponry. Whether remotely controlled or automatically programmed, KLO2 can be utilized to employ drones smaller than flies. Such small-sized drones were allegedly being used to deliver lab-cloned viruses or toxic drugs instrumental for spreading plague, virus, and disease; [and to] track spy satellites or large scale and hidden weaponries. KL02 could also be injected to devices like Google Glass to render the device obsolete or have the device controlled from 'someone' aside from its owner. The chip is also significant in making plausible Pentagon's ultimate dream of human-controlled robotic warrior. It can be injected to bionic prosthetics to control robotic nerves and limbs. 'Killing' the creators of this chip will prevent any leakage or selling of the technology to the Chinese and outside the Malaysian government. It is widely known that China is in its way of advancing its military-applications technology for warfare.[423]

Put it all together and what have you got? Well, that's not so easy to say, actually. However, while the above theory is strong on motive, it does little to explain the other necessary components of criminal/crisis resolution: means and opportunity.

"The failure of the air and sea search to find any wreckage around MH370's last known location, suggests a bomb did *not* bring down the plane."[424] There are very few ways that sabotage could have been stealthily accomplished. Although one way will be explored in the following chapter.

Seventeen

Explanation #8—Advanced Technologies

If you look around at the major stories on the Internet regarding the technology required to control a large aircraft remotely, you'll probably leave with the notion that it's a false rumor that it can actually be accomplished with a large plane. Well—drum roll, please! That is what would politely be called a "misdirection." Because, after taking a good look around at all the information available, it very clearly *is* possible. In fact, this new technology has apparently been around for quite some time. For example, the United States Navy not only admits to its existence, it even *advertises* it! And it's not just aircrafts that can be remotely controlled—there are also "autonomous boats" that travel without pilots or passengers. And we are not talking about little "drone boats" here either— these are large interceptor crafts, controlled by computers, that even travel in groups, acting in unison to "eliminate a threat" with their weapons systems when the case calls for it.[425]

You can even watch that technology demonstrated in a video—the Navy has apparently taken to network television to show off its new wares. The following are the verbatim words of Chief of Research, US Navy Rear Admiral Matthew Klunder, describing how simple it is to use a small iPad-like device to remotely take over the controls of a large helicopter, change its course, and land it where you choose to: "Just tap your button. Literally one touch of this tablet—the demonstration had no one in the cockpit, and safely landed that helicopter … Just with the one touch of the tablet."[426] That was done—and *even demonstrated for network television*—by two marines who had only been given "fifteen-to-thirty minutes max training on the pad."[427] Watch the video at news.yahoo.com/blogs/power-players-abc-news/technologies-once-available-only-in-movies-are-now-a-reality-for-the-us-navy-225652354.html.

For aircrafts, the technology was apparently available in 1996, but was rushed into the development of more sophisticated systems after the 9/11 terrorist attacks in 2001.[428] So first of all, let's not be naive. No matter how many "talking heads on the boob tube" reassure you that it is not possible to remotely control an airplane—*believe* us on this one—it *is* possible, and apparently has been possible for quite some time.

The following was published in the *Homeland Security News Wire* on December 4, 2006. The article is entitled "Boeing wins patent on uninterruptible autopilot system":[429]

"New technology can be activated by the pilots, government agencies, even on-board sensors; not even a tortured pilot can give up control; dedicated electrical circuits ensure the system's total independence ... takes control of the airplane away from the pilots and flies it to a predetermined landing position."[430]

For an excellent history on the development of this technology, see the *21st Century Wire* article by Shawn Helton, "FLIGHT CONTROL: Boeing's 'Uninterruptible Autopilot System,' Drones & Remote Hijacking," http://21stcentury-wire.com/2014/08/07/flight-control-boeings-uninterruptible-autopilot-system-drones-remote-hijacking/

Here's your bottom line, "The Boeing 777 along with other Boeing models, can in fact be flown remotely through the use of independent embedded software and satellite communication."[431]

BOEING HONEYWELL UNINTERRUPTIBLE AUTOPILOT

The Boeing Honeywell Uninterruptible Autopilot, or BHUA, is a patented software system in the Boeing-Honeywell AIMS (Aircraft Information Management System) in post-1995 Boeing aircrafts to overcome emergency situations, such as an in-flight hijacking, by overtaking control of the aircraft remotely for automated flight and landing. If BHUA is invoked in an emergency situation, it cuts all power from the cabin and cockpit and does not require, or allow, any pilot input to fly and land the aircraft. Control of the plane is taken over remotely via an up-linked flight plan which is transmitted to the aircraft's AIMS. In the event of emergency circumstances which necessitate the initiation of the Boeing Honeywell Uninterruptible

Autopilot, electrical power is removed from the flight deck and pilot authority passed to the navigational computer and auto-pilot system for automated landing.[432]

In plainer language: Control of the plane can be taken over remotely in an emergency situation, overriding the controls in the cockpit, to fly and land the plane. In fact, both of the huge corporations involved in developing and implementing this technology—Boeing and Honeywell—have actually publicly acknowledged that BHUA exists. Boeing issued a statement to reporters, documented on March 3, 2007, which acknowledged the existence of the Boeing Honeywell Uninterruptible Autopilot. That statement was apparently in response to Civil Case Number 3:07cv24 in federal proceedings in the District Court of North Dakota, which concerned the existence of the new avionics.[433] The Boeing spokesperson stated:

"Once this system is initiated, no one on board is capable of controlling the flight, making it useless for anyone to threaten violence in order to gain control."[434]

Boeing had previously even patented the remote control applications of the software system. On February 19, 2003, Boeing Corporation filed patent number US7142971B2, entitled "System and method for automatically controlling a path of travel of a vehicle."[435] It is referred to in the patent as "engaging an automatic control system" that "may be automatic or manual from inside the vehicle or remotely via a communication link ... The control commands may be received from a remote location."[436] On April 16, 2003, Honeywell filed patent number US7475851B2, titled "Method and apparatus for preventing an unauthorized flight of an aircraft."[437] It is referred to in the patent as a system "for controlling the aircraft" and "to carry out a transfer of control from the FBW (the 777's fly-by-wire avionics system) from the cockpit controls to the AFCS (automatic flight control system)."[438]

Therefore, to say that such a system does not exist, clearly stands contrary to the known facts.

FAA Concerns, Expert Opinions, and Further Demonstrated Technology

The Federal Aviation Administration apparently had serious concerns about the special modifications to the avionics systems of the Boeing 777. The FAA issued a "Special Conditions Action"—a safety warning due to a deviation from existing or proposed guidelines—specifically on the Boeing 777. It stated that the aircraft, "as modified by the Boeing Company," regarding its "novel or unusual design features" to its "on board network systems" and "on board computer network system," that were not part of Boeing's original guidance material, "may enable the exploitation of network security vulnerabilities and increased risks potentially resulting in unsafe conditions for the airplanes and occupants." The FAA special action was entitled "Special Conditions: Boeing 777; Aircraft Electronic System Security Protection From Unauthorized Internal Access." It reads:[439]

> This proposed data network and design integration may result in security vulnerabilities from intentional or unintentional corruption of data and systems critical to the safety and maintenance of the airplane. The existing regulations and guidance material did not anticipate this type of system architecture or electronic access to aircraft systems. Furthermore, regulations and current system safety assessment policy and techniques do not address potential

security vulnerabilities, which could be caused by unauthorized access to aircraft data buses and servers.[440]

As one aviation security consultant—Jim Termini of Redline
Aviation Security—summarized it, when asked about the possibility that Malaysia Flight 370 was remotely "cyber jacked":

"I've seen evidence from an aircraft manufacturer, not too
long ago, that would indicate it *is* possible. And I believe that
it's something that the agencies within the region and Boeing
must be looking at very seriously."[441]

Dr. Sally Leivesley, PhD, is a former consultant to the
British Home Office (the UK equivalent of the US State
Department) Scientific Research and Development Branch and
has an extremely impressive track record which you can access
here, www.newrisk.com/drsallyleivesley.html. Dr. Leivesley's
opinions should be taken quite seriously. Here's what she had
to say about what may have happened to Flight 370:

"It might well be the world's first cyber hijack. There
appears to be an element of planning from someone with a very
sophisticated systems engineering understanding. This is a very
early version of what I would call a smart plane, a fly-by-wire
aircraft controlled by electronic signals."[442]

Well, that's some pretty serious stuff, and from a very serious analyst who is regularly consulted on important matters,
from Ebola to cyber security:

"Dr. Leivesley, who now prepares businesses and governments for potential terrorist attacks, added that once the plane
is air-side, you can insert a set of commands and codes which
can begin a new set of processes."[443]

It is even put forward by Dr. Leivesley that a remote control hijacking of Flight 370 could have been accomplished via use of a cell phone.[444]

There is ample evidence that the capability of remotely controlling an aircraft existed even *before* the admission and patent registrations of Boeing and Honeywell. In 1996, remote control technology was tested and demonstrated. One of the witnesses to that demonstration was Avionics Technician Wayne Anderson.[445] Anderson documented that successful use of remote control technology in an interview with Rob Balsamo, which can be accessed online at http://pilotsfor911truth.org/ Remote_Control_Whistleblower.html.

Therefore, remote control technology has apparently been around for quite some time.

Andreas von Bülow has a distinguished career in government, as both a leader and statesman. His job titles have included:

- Minister of Defense, Germany (cabinet-level position)
- Minister for Research and Technology, Germany (cabinet-level position)
- Member, German Parliament (twenty-five years), including the Committee on Intelligence Services, which oversees and supervises the national intelligence agencies and has access to classified intelligence information.

Von Bülow has also authored several books on intelligence agencies, one of which is titled *Die CIA und der 11. September. International Terror und die Rolle der Geheimdienste* ("The CIA and September 11: International Terror and the Role of Intelligence"); a very interesting book that is conveniently unavailable in English.[446] He firmly believes that remote control technology exists and even goes a step further, stating that it was used during the 9/11 attack in New York City to direct the hijacked planes into their targets.[447]

The history of the development of the Boeing Honeywell Uninterruptible Autopilot system is documented on Wikipedia, http://en.wikipedia.org/wiki/Boeing_Honeywell_Uninterruptible_Autopilot.

Need more proof? How about this one? NASA proudly demonstrated remote controlling a Boeing 720 way back in 1984: A test was conducted at Edwards Air Force Base, filmed, and was successful, as the remote controlled jumbo jet was crashed intentionally in a "Controlled Impact Demonstration":[448]

"During the fourteen flights, the Boeing 720 was controlled remotely by a pilot at a ground-based console for sixteen hours and twenty-two minutes, including ten takeoffs, sixty-nine controlled landing approaches, and thirteen landings on the abort runway."[449]

You can even *watch* that remote-controlled crash online. Just go to www.youtube.com/watch?v=3xEiPvLWEeo.

So the fact that it is even a publicly contested point, with otherwise-intelligent people implying that the technology does not really exist, is yet further indication that, for whatever reasons, we are not being told the complete truth about things. It also suggests that our government is probably the least likely source to divulge that truth too. Basically, if we ever thought we actually knew what's really going on—we'd better think again. Technologies are now so sophisticated that it boggles the mind. There's more.

In August of 2001, this technology was further demonstrated by Raytheon, which successfully took off and landed a Boeing 727 six times at Holloman AFB in New Mexico without a pilot on board. Raytheon also developed a sensor suite for the Air Force's Global Hawk drones and Raytheon Network Centric Systems has recently won multiple contracts to help develop advanced communications systems for the E-4B, the US government's so-called doomsday plane that was spotted above the White House shortly before the strike on the Pentagon and which has since been confirmed was one of four functioning Doomsday Planes operating in the skies on that day.[450]

The specially modified E-4Bs—which are designated "Advanced Airborne Command Posts"—are designed to become the operational control center for the United States Government in the event of a catastrophe.[451]

You can see the so-called "Doomsday Plane" right here, www.youtube.com/watch?v=SMK5bmdAEHc. *And,* you can watch Anderson Cooper and CNN's John King talking about and even showing us the Doomsday Plane that officially wasn't there, in the following report, at the link below: "Ask the Pentagon, and it insists, this is not a military aircraft. And there is no mention of it in the Official Report of the 9-11 Commission."[452] Yet, *by golly, there it is*—right above the White House, in the world's most highly restricted airspace. You *have* to see this one: www.youtube.com/watch?v=SMK5bmdAEHc.

Officially, the government denies that the huge plane above the White House during 9/11 was military or, for that matter, that it was even *there,* because, officially Doomsday Planes do not exist.[453] Even though it has been established beyond question that they *do exist—much* like remote control technology. And, oh by the way—if you guessed that the special E-4B Advanced Airborne Command Post aircraft is manufactured by Boeing Corporation, you are correct.[454]

Eighteen

Explanation #9—US Military Air Base, Diego Garcia

If you've ever wondered about the secret US military base known as Diego Garcia, all we can say is: *join* the club. A lot of people have wondered about it. And you can *keep wondering* too, because The Powers That Be are very hush-hush about that one.[455] An Article in *The Star* says it well:

> As a US military and intelligence base, the island has remained under a cloud of secrecy for decades ... the base was a little-known launch pad for the US wars in Iraq and Afghanistan and may house a top-secret CIA prison where terror suspects are interrogated and tortured.[456]

There is a classic book that everyone should read, and hardly anyone will—it's out of print and they don't *want* it to be read—called *Island of Shame: The Secret History of the U.S. Military Base on Diego Garcia*, by David Vine. It details how "imperialism, military prerogative, and racism have all combined to deny a people a home simply because they were in the way."[457]

Diego Garcia is a small island south of India. It's a very strategic location because it is one from which fighter aircrafts can quickly reach anywhere from Africa to northern Asia. Our British friends technically lease us the land for the base. The Brits got it—by force—during the "good old days" of great colonial expansion at the height of the British Empire, as it's so fondly known.

The base was reportedly also used as a "black site" for extraordinary renditions.[458] For those of you who don't know, extraordinary rendition is the term for the illegal kidnap and torture of any persons even suspected of terrorism, made technically "legal" nowadays as a result of the very ill-named *Patriot Act*, the purpose of which was to make us all feel safer— whether we actually are or not. No proof needs to be produced in order for a rendition to take place—a mere accusal will do just fine these days. They just burst into some family's home in the middle of the night, throw a sack over some guy's head, and toss him in a cell where they get to do stuff like "sensory deprivation" to him. It's like in the movies, only it's real. You just won't hear much about it on the news.

Speculation was forced to widen when it was disclosed that all the evidence pointed away from a catastrophic event that quickly brought down Flight 370:

"The failure of the air and sea search to find any wreckage around MH370's last known location, suggested a bomb had *not* brought down the plane."[459]

Mainstream media may have ridiculed and attempted to marginalize some of the "less mainstream" theories about Flight 370 but—given the facts—who the hell can blame them?

CONCLUSION: FLIGHT 370 DID NOT EXPLODE—IT VANISHED

A well-written article in *Natural News* states: "The inescapable conclusion from what we know so far is that Flight 370 seems to have utterly and inexplicably vanished. It clearly was not hijacked (unless there is a cover-up regarding the radar data), and we can all be increasingly confident by the hour that this was not a mid-air explosion (unless debris suddenly turns up that they've somehow missed all along). The inescapable conclusion is that Flight 370 simply vanished in some way that we do not yet understand. This is what is currently giving rise to all sorts of bizarre-sounding theories across the net."[460]

In other words, the only thing that the authorities seem to be sure of is that a Boeing 777 vanished—and *planes simply don't vanish* in this age of modern technology. Hence, the theorizing began, and you can hardly blame anyone for that. That's perfectly logical—even prudent! The fact that it was then announced that Flight 370 made a sharp westerly turn when it went off course, heading into the middle of the Indian Ocean, for no apparent reason—raised red flags in some minds when that course was determined to be for the base at Diego Garcia. As one researcher quickly noted:

"**What about Diego Garcia?** ... There's 'nothing' west of Kuala Lumpur and Indonesia except the huge US naval base at Diego Garcia, in the middle of the Indian Ocean. Diego Garcia appears to fall near, but inside, the limits of the plane's maximum travel distance. That raises a few questions:

a. **Could Diego Garcia have been the target** of a terrorist plot using this plane? Could you imagine a US naval base having to choose whether to shoot down a fully-loaded civilian airliner out of the sky? There was a lot of discussion about how there were no 'rich targets' in that area. Diego Garcia is one such rich target.

b. **What radar coverage does Diego Garcia have**, and has it been checked?

c. **What about our ships and subs in the area**—would their radar and sonar potentially have picked up a plane heading in their direction or crashing in the seas?

d. **Satellite coverage.** I just saw an expert on CNN say there's no reason for there to have been satellite coverage over the Indian Ocean, except that Diego Garcia is there. So are we sure there's no satellite coverage?"[461]

Wouldn't it be extremely naive to suggest that there was no radar and satellite coverage for one of the most strategic US military bases on the entire planet? Wouldn't they have been *extremely* interested in the development of a *way* off course commercial jetliner that had suddenly gone off-program and was heading their way? Do they actually think we're so naive that we believe there was no radar *or* satellite tracking of a development that dramatic? And why won't they tell us? What are they hiding?

"How, in this high-tech age of uber-surveillance, in which hundreds of satellites sweep the earth and modern aircraft have multiple communications systems with triple redundancies, can a plane vanish?"[462]

It had even been announced that Malaysian authorities had at one point made an amazing discovery: the flight simulator at the home of the pilot of Flight 370 contained a program for the landing strip at the base at Diego Garcia, apparently for the purpose of practicing the landing procedure there. Atop that, that program had been deleted from the flight simulator, but was still apparently visible on its hard drive.[463] To some, that seemed rather ominous. The runway at Diego Garcia is long enough for a Boeing 777:

"Today it was also revealed that a remote island in the middle of the Indian Ocean with a runway long enough to land a Boeing 777 was programmed into the home flight simulator of the pilot of the missing Malaysia Airlines plane."[464]

Also, an official denial came from the authorities—the United States, in this case—vehemently, and very specifically, stating that the plane *did not land at Diego Garcia*.[465] Oddly enough, that denial seemed to draw more attention to the possibility; it almost seemed like the authorities were a little too anxious about that possibility:

"It has also been stated that Malaysian authorities have denied reports of the plane landing at the United States military base in Diego Garcia located in the Indian Ocean, fueling even more suspicion, as the plane is said to have been heading on course toward Diego Garcia with its last listed radar contact near Silver Island. Has the public been deliberately steered away from other possible locations of the plane, such as areas within or near the Indian Ocean?"[466]

Lest we forget, some of the more positive eyewitness testimonies about possible Flight 370 sightings came from an area that would also line up with a flight to Diego Garcia:

"The investigation into the Diego Garcia, an overseas territory of the UK, which is rented to the US and is now a huge American naval base follows fresh eyewitness accounts of a 'low flying jumbo jet' being spotted in the Maldives. People on the island of Kuda Huvadhoo reported seeing a plane on the morning of the disappearance of Malaysian Airlines flight MH370, according to *Haveeru*, a news website in the Maldives.

"Islanders said a white aircraft with red stripes across it—which would match the missing plane—was seen traveling north to south-east towards Addu, the southern tip of the Maldives. An eyewitness told the website: 'I've never seen a jet flying so low over our island before. We've seen seaplanes, but I'm sure that this was not one of those. I could even make out the doors on the plane clearly.

"'It's not just me either, several other residents have reported seeing the exact same thing. Some people got out of their houses to see what was causing the tremendous noise too.'"[467]

And then, in an even *more* dramatic development, a journalist asserted that information from one of the passenger's cell phones was traced to the GPS location of, *you guessed it*—Diego Garcia. Dovetailing with the reports about the pilot practicing landing at Diego Garcia was a report by freelance journalist Jim Stone stating that a photograph sent from the iPhone 5 cell phone of a Flight 370 passenger was verified by its technical data as originating at the GPS coordinates for Diego Garcia:

"Amazingly, Stone claims that metadata within the photo yields evidence confirming '100 percent' that Phillip Wood sent the photo, along with a brief voice activated text, from GPS coordinates which put Wood only a few miles away from the US controlled Diego Garcia military base which is located on an island south of the Maldives in the Indian Ocean."[468]

Even the US military got in on the act—with a three-star General, no less. Retired Air Force Lieutenant General Thomas McInerney—in a nationally televised interview—said that the authorities, including the US government, "has not come clean with this," insofar as divulging what it knows about Flight 370.[469] The general added that it was clear that the plane was hijacked. He also chimed in that, after studying the facts, in his professional opinion, the plane had been stolen. The interview is still available online, http://video.foxnews.com/v/3363882454001/gen-mcinerney-us-hasnt-come-clean-with-flight-370-data/#sp=show-clips.

"Someone wanted that airplane. You're not going to fly into the Indian Ocean to crash it. You'd have crashed it in the Gulf of Thailand if they wanted to do it. Someone wants to use that airplane for something."[470]

Pretty amazing stuff, huh? The general continued:

"I don't believe that this airplane crashed. I do believe it was hijacked. I do believe it landed someplace … But something very nefarious has happened."[471]

In fairness, it should be added that General McInerney stated that he believed that Pakistan was the probable destination for the flight. One report on *Fox News* said that a source told them that Boeing believes the plane was in Pakistan.[472] Malaysia was also asking Pakistan for help to find the missing plane.[473]

"And they're [the government of Malaysia] not coming clean with this. And the US government has not come clean with this. And Boeing and Rolls-Royce [the manufacturers of the jet engines, which transmit information back to the company automatically] have a lot of data that they haven't put out in the public airways, and I think they ought to."[474]

All in all, the evidence indicates that diverting Flight 370 to Diego Garcia, by whatever means, for whatever reasons, cannot be ruled out as a possibility. And we're not saying that we know what those reasons are, or that that is definitively what happened to Flight 370. But we *are* saying that diversion to Diego Garcia cannot currently be eliminated as a possibility.

Nineteen

Conclusions

O ur meta-analysis of the evidence has uncovered some very important information that you probably were not familiar with before you picked up this book:

- A distress call from Flight 370 apparently *was* made, contrary to the official version that there was not one.[475]
- The flight crews of two commercial airliners flying near Flight 370 were asked to radio-communicate to the plane and did so, receiving only a garbled response.[476]
- Fighter planes *were* reportedly sent up to identify and escort Flight 370, though official reports deny that.[477]
- There are numerous eyewitness reports of a low-flying, burning jetliner, matching the description of Flight 370, near its known locations.[478]

- Malaysian police admitted that they faked the photographs of the two Iranian passengers traveling with stolen passports.[479]
- Radar information was also incorrect. "Flight 370 did not dramatically climb to 45,000 and then dive below 23,000 feet after completing a U-turn before it disappeared."[480]
- Sophisticated listening stations using infrasonic detection can detect the sound of a plane crashing, but reported that nothing was detected insofar as Flight 370 is concerned.[481]
- The number of cell phone pings from a large jetliner "would be a chorus" to the hi-tech listening stations the United States has around the world, said a technology expert, who was amazed that the plane could not be located via those digitized signals.[482]
- Cell phones should have functioned at some of the reported locations and, indeed, the copilot *did* place a call with his cell phone.[483]
- Another cell phone message was reportedly traced to GPS coordinates near the Diego Garcia military base, via the software embedded in the message.[484]
- Twenty of the passengers on the flight were employees of Freescale Semiconductor, a company that reportedly had developed a new microchip that "might be the most potent next-generation war weaponry."[485]
- Contrary to mainstream media misinformation, taking control of an aircraft remotely, is a well-documented capability. It quite clearly *can* be done.[486]
- Contrary to mainstream media misinformation, there is no event in all of aviation history that even comes *close* to

the circumstances of Flight 370. Air France 447, though often compared, was quickly located, and was precisely where they thought it would be.

- Contrary to mainstream media misinfo, commercial jetliners *are* tracked everywhere, even in remote areas of ocean: "The aircraft's flight computer, programmed to send automatic position reports by satellite, continues to transmit data."[487]
- A startlingly dichotomous difference was evident with the Boeing 777 of Flight 17, compared to the Boeing 777 of Flight 370. In the case of Flight 17, US Intelligence immediately revealed known and precise data. In the case of 370, there was a notable *retention* of data—they weren't even sure what ocean they should be looking for it in.

The above points are detailed extensively in the earlier chapters, "Anomalies," "Comparison," "Advanced Technologies," and "Sabotage." Where does that evidence lead us? Even CNN conceded that initial reports were very faulty; as a CNN anchor said, regarding those altitude changes initially reported: "Remember those reports? Well, apparently, they were dead wrong."[488]

The effect of turning off the transponder is to make the aircraft inert to secondary radar, so civil controllers cannot identify it. Secondary radar interrogates the transponder and gets information about the plane's identity, speed, and height. It would however still be visible to primary radar, which is used by militaries.[489]

When Malaysian authorities pointed to a probability of hijacking, it gave birth to the theory that the plane made a sharp left turn. That was presented by authorities as the new information we were to accept.[490] Suddenly, everything pointed to hijacking: a Malaysian General said—*days* after the incident—that military radar showed the plane at a westerly location; the prime minister said the plane's systems were *gradually* turned off and that such actions were "consistent with deliberate action by someone on the plane."[491] To many observers, however, that raised some disturbing questions. How, why, and *when* would that course change have been programmed into the plane's flight computer if, as it seemed, there had been an emergency in the cockpit? And if the plane had veered sharply off course, suddenly changing in mid-flight, from its northeast flight pattern to a new course that was practically due west, wouldn't the passengers have noticed that, wondered what was going on, and attempted to make a phone call, even from a satellite phone? Then the *New York Times* came to the rescue with an explanation of how those course changes were all pre-programmed into the flight computer. They even explained that nobody would have noticed and, somehow—though no one knows quite how—they even divined that the turn was at a nice, easy twenty degrees that none of the passengers would really even have noticed:[492]

"Although it was later confirmed that the last ACARS transmission showed nothing unusual and a normal routing all the way to Beijing, *The New York Times* reported 'senior American officials' saying on March 17 that the scheduled flight path was pre-programmed to unspecified western coordinates through the flight management system before the ACARS stopped functioning, and a new waypoint 'far off the path to Beijing' was added. Such a reprogramming would have resulted in a banked turn at a comfortable angle of around twenty degrees that would not have caused undue concern for passengers."[493]

And if you're wondering how anybody could actually know such a thing, then we've got the answer to that one:

They couldn't. But note this: that miraculous explanation also conveniently explained, albeit very weakly, why there were no phone calls attempted from the plane. In the "Anomalies" chapter, we provided the calculations of an aviation consultant, Simon Gunson, which, he believes, prove that the aircraft never made that sharp westerly turn. That information can be found in the section, "Flight Path Change." As that consultant concludes on the matter:

"The *New York Times* is wrong. Nobody typed new coordinates into the CPDLC [Controller Pilot Data Link Communications], to give the keypad its correct name, because MH370 never turned west from IGARI.[494] There was a report, however, that the US Navy listening station at Rayong U-Tapao [a joint civil-military airport in Thailand] heard a distress call from MH370 in which pilots declared an emergency because their cabin was disintegrating and that they would make an immediate landing.[495]

Other knowledgeable researchers have shared those same suspicions about the "late-arriving" satellite data. Jeff Wise, in an article entitled "Why Inmarsat's MH370 Report is a Smokescreen," cited specific discrepancies in the reported satellite data.[496] Other sources have also confirmed the distress call that was reportedly picked up and recorded by the US Navy listening post in Thailand:

"The *China Times* reported on 8th March that MH370 made a distress call that they would make an immediate landing because the 'cabin faced disintegration.' The distress call was recorded by a US Navy listening post VTBU-Rayong at U-Tapao, Thailand. The US Embassy handed over tape recordings to the Malaysian government, but your government will not mention the distress call in the preliminary accident report."[497]

In a statement appropriately titled "Evidence of a Massive Hoax by the Malaysian Government," Simon Gunson further explained the controversy surrounding that distress call:

"A veil of secrecy has been drawn over these distress calls. A US Navy vessel, the USS Pinckney, seems to have been the only USN asset within VHF radio range under the flight path off Vietnam. If you check out the Facebook page for USS Pinckney circa 9th March, Chief Mate Edmunds of USS Pinckney ordered families of crew not to discuss what they knew or had heard from their loved ones and imposed a security blanket ordering families not to disclose the location of USS Pinckney. The NSA also declined requests to divulge what the US Government knew, citing Presidential Executive Order #13526."[498]

That would seem to make a lot more sense than what has been reported, that no one heard anything. As to why it has not been reported, one can only speculate. But, as Gunson continued, that would indicate the actual emergency aboard Flight 370 and that "it kept flying by autopilot after a catastrophe off Vietnam."[499]

The resulting enigma of all the evidence is that:

Only a catastrophic event could have caused a global loss of all emergency methods, systems, and equipment.

Yet—if the official reports are true—the plane changed course and continued to fly for several hours—which should have allowed <u>ample time to declare an emergency</u>, send emergency signals, and allow ground tracking and location identification via radar and satellite systems.

In other words, both possibilities seem to cancel each other out. They are what could be termed *mutually exclusive evidence.* Neither conclusion seems possible because many of the premises should not have occurred if the conclusion is correct. If the flight changed course—due to *any* non-catastrophic problem—why did the flight crew fail to declare an emergency? Why didn't they at least send an emergency squawk to air traffic control? That only takes half of one second. Why did the emergency locator transmitters and a myriad of other high-tech fail-safe security and communications measures immediately cease to function?

Alternatively, if the flight suffered a catastrophic event, like being struck by a missile or facing a similar-level crisis, why was the course apparently altered by the pilots? How could they or the plane have even functioned? And why wasn't the plane trackable by radar or satellite?

The experts have failed to come up with a believable explanation for these *many* anomalies. The fact is that there are now very few possibilities that even come remotely close to plausibly explaining all the events concerning Flight 370.

Concentrate for a moment, on the broader aspects of this whole crazy scenario: In a post-9/11 world, a huge commercial airliner—*without* advising air traffic control—turns sharply off its northeast flight pattern, heading back south or southwest. It is impossible to believe that some form of radar does not track such an extreme move. As far as answers to the question, "What do they do when they see that?"—*nothing* is not one of them. Protocol, based on historical precedent, is that they immediately assume a threat, scramble combat aircrafts, and a few minutes later, there are two fighter jets off each wing of the rogue plane; F-16s, or whatever the best that their military can put up there at the time. They get so close that the fighter pilots can literally see inside the cockpit and determine status. They inform the crew that they are escorting the plane to a landing and to follow them and not alter their course, under threat of being shot down if they refuse. That's protocol for that event, in a post 9/11 world, especially if the plane is headed toward the Petronas Twin Towers in your capital city.

Since Flight 370 is considered such a baffling case, we remembered the wise words of FBI veteran, Zack Shelton, who assisted us greatly with our investigation of the JFK assassination in our first book, *Dead Wrong*. Zack's credo, as far as how to follow the evidence in an investigation, is this: "I don't have any theories. All I have are the facts."[500]

Invoking that reasoning in this case: What do we really know about what happened to Flight 370?

- We know that—reportedly—no debris has been found, and that there *should* have been a lot of debris, even if it crashed in the ocean. A ditch of a commercial airliner on the ocean is usually very messy and, even with moderate success, creates huge pieces of debris, many of which float. Some debris should have been found by now if it crashed at sea—yet none has. And that means one of two things: either the debris was recovered surreptitiously, or the plane did not crash in the ocean.

- We know we are told that radar didn't track the flight and that, normally, radar *tracks all the flights*.

- We know that—officially, at least—no distress call of any type was issued by the flight crew and that, normally, declaring an emergency is one of the first things that they do.

- We know from the last communications that everything was absolutely normal, and then, a minute later, all communications were lost, and radar tracking supposedly was too.

- We know that, when Vietnam air traffic control did not pick up the flight as expected on their radar, they had the crews of two nearby commercial airliners attempt to contact Flight 370 via radio, and that they *did* receive a radio response from 370 but it was garbled and unintelligible.
- We know that Vietnam air traffic control then notified Kuala Lumpur ATC that Flight 370 must have turned back. When Kuala Lumpur ATC did not respond, a few minutes later, they *again* notified KL-ATC.
- We know that—officially—fighter aircrafts were never sent up to intercept and escort Flight 370—and that it should have been standard protocol for Malaysia authorities to do that.
- We know that the aircraft, for some reason, reportedly maintained radio silence.
- We know that there was an incredibly odd four-hour gap between the time that Vietnam ATC alerted KL-ATC, *twice*, that 370 must have turned back—and the time that Malaysia initiated search and rescue efforts.[501]
- We know we have not been given satisfactory explanations, and that this is an odd case, where many suspect that things are being hidden.

Given all of this, we have to look at how each of the possibilities matches up with those *known facts*—and most of them don't match up very well. Let's look at them all:

- Weather: there is no indication that weather caused any problems.
- Hijacking: no indications of it, as far as a typical skyjacking, and there should have been some evidence of that by now.

- Pilot Suicide or Pilot Error: no evidence of this, or real motive for it.
- Aliens/Other Extraordinary Means: no evidence of it.
- Sabotage: a bomb would have left substantial debris. Where is the crash site?
- Missile Strike: radar would show that. Where is the debris? Why did the plane keep flying?
- Advanced Technologies: possible, but by whom and why? And where's the plane?
- Fire or Mechanical Failure: possible, but a fire would have had to be big enough to *not* give the crew any time to communicate their emergency; yet *small* enough to be extinguished by the aircraft's systems and then continue flying.

To put it as simply as possible:

Only a catastrophic event could cause the simultaneous global failure of all safety and emergency systems plus back-up fail-safe emergency systems, on board and in communication and tracking with, a Boeing 777. Many types of catastrophic events can be ruled out. Those that remain possible are:

- **Massive Fire** (only partly plausible because there still should have been time to declare an emergency)
- **Missile Shoot-Down** (accidental or intentional, with surreptitious debris retrieval)
- **Hijacking/Advanced Technologies** (such as a "cloaking technique" of electronic warfare)

That's why some investigators have speculated that it was an event involving new technologies that somehow "cyberjacked" the plane remotely; or a failed or thwarted attempt at a cyber hijacking. The technology exists,[502] even though the illustrious

Powers That Be keep politely telling us that the technology does *not* exist; reminding one of that prophetic warning from Ralph Waldo Emerson, "The louder he talked of his honor, the faster we counted our spoons."[503]

Cyber hijacking is about the only possibility that fits the above circumstances insofar as the known evidence regarding the actions of the plane. We're not saying that's what happened. The evidence, at this juncture, is inconclusive. We *are* saying that the official version of "We lost the plane and it may never be found" is an obvious ruse and a very weak one at that. What we keep hearing as the official explanation of this case has got more holes in it than Swiss cheese—and a pretty "holey" batch of Swiss cheese. Here's how one so-called expert rationalized the inability of modern technology to locate the aircraft:

"I don't think people realize just how big the world actually is."[504]

Wow! That's quite an excuse. We hadn't heard *that* one before. But she was right—the search area just got bigger and bigger until, at one point, it covered 2.96 million square miles.[505] Which is pretty darn incredible, considering that—lest we forget—we live in an era where it's common knowledge that a simple cell phone can be located via GPS, just about anywhere on the planet. So, forgive us, but some of us were a bit surprised that they said they managed to actually lose a Boeing 777.

Believe it or not, the documentary by the Smithsonian on Flight 370 actually concluded with a crash investigator suggesting that maybe they should put a simple GPS device on commercial aircrafts so that they can track them better.[506] Gee, *really?*—You *think?* And then, the narrator of the documentary, stated dramatically that: "It may take *years* for these new technologies to become accepted and implemented. And even then, none will help solve the enduring mystery of Malaysia 370."[507] *What* new technology? We're talking about *GPS*. It's been around for years. Not very reassuring, is it? *Or,* maybe somebody could lend them a cell phone. The whole thing just makes you want to say "Beam me up, Scotty, there's no intelligent life here." In fact, the person who suggested that planes carry a GPS device was the same crash investigator who made the remark that "I don't think people realize just how big the world actually is."[508] Maybe somebody should bring her into the War Room at the Pentagon, so she can see just how *small* they think the world has gotten. And then they could explain to her how simple GPS tracking is. Take my cellphone—*please!*

We concur with the conclusion of another veteran researcher, Dean Garrison over at *D.C. Clothesline*, which prides itself on "Airing out America's dirty laundry":

> Again, as I have said before, the one thing that I believe about Flight 370 is that, in 2014, it is ridiculous that this plane could disappear without someone knowing where it landed or crashed.[509]

Matthias Chang is a prominent attorney in Malaysia, and former adviser and political secretary to former Malaysia Prime Minister Dr. Mahathir Mohamad, and wrote the following:

> [The] MH370 "disappearance" took place during two major military exercises in Thailand. NSA can track your phone calls so long as you are using your phone and as long as your phone is switched on, you can also be located. You cannot hide from the intelligence apparatus. This is from Edward Snowden. An airplane runs on

electronics. It has miles of electrical wirings that control and transmit data that enable the plane to fly. So, even if the transponder is disabled and if the plane is still flying as in the case of MH370, then there would be electrical signals emitting from the plane. And as long as the plane is flying, its engines (Rolls-Royce fitted with FADEC) will of course be running and signals would be emitted to indicate either the engines are running efficiently or there could be malfunctions, so that the pilot can take remedial actions. In the case of Rolls-Royce engines, the HQ monitors all engines while in flight.[510]

That's a recurring theme. As an article in the *21st Century Wire* reads, "Air traffic controllers would have known the exact moment that something had changed during the course of this flight."[511]

It also pays to keep in mind the "unseen story" here. If, as one article suggested, the Petronas Towers—known as the Twin Towers of Asia—were the real target of whatever actually transpired with Flight 370, that might explain a lot of the strangeness surrounding the whole story.[512] A surprising number of very public people seem to have arrived at the conclusion that Flight 370 was "taken" by some nefarious means, such as cyberjacking.[513]

One of the first articles we read after Flight 370's disappearance seemed the best at concisely putting together the basic overriding points of the case. It reads:

> While investigators are baffled, the mainstream media isn't telling you the whole story, either ...

- Fact #1: All Boeing 777 commercial jets are equipped with black box recorders that can survive any on-board explosion
- Fact #2: All black box recorders transmit locator signals for at least 30 days after falling into the ocean
- Fact #3: Many parts of destroyed aircraft are naturally buoyant and will float in water
- Fact #4: If a missile destroyed Flight 370, the missile would have left a radar signature
- Fact #5: The location of the aircraft when it vanished is not a mystery
- Fact #6: If Flight 370 was hijacked, it would not have vanished from radar. Conclusion: Flight 370 did not explode; it vanished ... The inescapable conclusion is that Flight 370 simply vanished in some way that we do not yet understand. [514]

One facet of a hijacking scenario that has not been the focus of much attention is the possibility that Flight 370 was actually part of a terror incident, 9/11 style, and that action was either foiled, staged, or something went wrong. In an interview with Dr. Kevin Barrett, a former university professor who is an outspoken critic of the War on Terror, that topic was breached:

> *Press TV*: Well, Dr. Barrett, let's look at it from a technical perspective. Do you think it is possible for no radar system to have picked up this airplane, even as it allegedly went way off course for hundreds of kilometers?
>
> *Dr. Barrett*: No, that is not possible. In fact this aircraft cannot be just lost. It cannot have just disappeared. This makes no sense. There is military radar and satellite

coverage of that area. The CIA base in Alice Springs, Australia knows precisely what happened to that plane. And it is interesting the Malaysian government has asked them and they are not getting any response.[515]

And here is another strong dose of common sense—one that echoes a note many have heard playing in the backs of their minds. The following is from a man who states that he worked for the NSA in a high position. It might sound a little "out there" to some—but it actually makes quite a bit more sense than the official version does, folks. So *here ya go!* It's reproduced exactly as it was written:

The NSA has black radiotelescopic satellites in orbit that have dishes in excess of 180 feet across, a feat made simple by the fact that the dishes do not need to be supported while in the weightlessness of orbit, so they don't need a lot of structural materials in them. These telescopes are so sensitive that if you are out in the country, in the wilderness away from all the other radio chatter and electromagnetic interference, a simple battery operated watch is all it will take to give away your position. And the circuits in a portable FM or AM radio are even better for blowing your cover.

You CANNOT HIDE FROM THE NSA NO MATTER WHAT absent abandoning all electronics, so I beg to question—HOW COULD THEY HAVE LOST FLIGHT 370 AND SUBSEQUENTLY NEVER FOUND IT WHEN THEY HAVE HUGE RADIOTELESCOPES IN SPACE THAT CAN DETECT THE OPERATION OF CIRCUITS AS LOW POWERED AS THOSE IN A WATCH?[516]

"Just journalists," you say? Okay. Fair enough. How about a couple of prime ministers? How about the CEO of the airline with the biggest fleet of Boeing 777s? Sir Timothy Charles Clark is President and CEO of Emirates Airlines, the largest airline in the Middle East.

> Clark, whose company owns the largest fleet of 777 aircrafts, numbering some 127 or more, said he "does not accept the findings from officials that not a trace of the commercial jet has been found. Moreover, he questions the computer and GPS (global positioning satellite) data."[517]

As Sir Clark noted:

> There hasn't been one overwater incident in the history of civil aviation—apart from Amelia Earhart in 1939—that has not been at least five or ten percent trackable. But MH370 has simply disappeared. For me, that raises a degree of suspicion. Our experience tells us that in water incidents, where the aircraft has gone down, there is always something. We have not seen a single thing that suggests categorically that this aircraft is where they say it is, apart from this so-called electronic satellite "handshake," which I question as well ... I'm totally dissatisfied with what has been coming out of all of this. Disabling it [the tracker] is no simple thing and our pilots are not trained to do so. But on flight MH370, this thing was somehow disabled, to the degree that the ground tracking capability was eliminated.[518]

Sir Clark also had some very interesting comments in October 2014:

My own view is that probably control was taken of that
airplane. It's anybody's guess who did what. We need to
know who was on the plane in the detail that obviously
some people do know. We need to know what was in the
hold of the aircraft. And we need to continue to press all
those who were involved in the analysis of what happened
for more information.[519]

That's pretty strong stuff from a guy who is in charge of 127
Boeing triple-sevens. It gets even better. He also said "it would
be unlikely that the aircraft's pilots would have been able to
disable all the jet's tracking equipment."[520] Sir Clark continues:

I do not subscribe to the view that the Boeing 777, which
is one of the most advanced in the world and has the most
advanced communication platforms, needs to be improved
with the introduction of some kind of additional tracking
system. MH370 should never have been allowed to enter a
non-trackable situation.[521]

If you think *that's* dramatic, get a load of this: for Sir Tim,
there is only one logical conclusion as to what happened to the
highly advanced and reliable passenger jet: "MH370 was, in
my opinion, under control, probably until the very end."[522]

Both the current and a former prime minister of Malaysia
have echoed those doubts as well. A *Daily Mail* article reads:

"Dr. Mahathir's blog posts come after the current Malaysian
prime minister, Najib Razak, described the location by satellite of purported MH370 debris in the Indian Ocean as
'bizarre' and 'hard to believe.' Najib told CNN [that] he

did not believe it when he first heard about the critical satellite data on which the current search in the Indian Ocean is based. 'To be honest, **I found it hard to believe,' said the prime minister.**"[523]

All things considered, especially from a historical standpoint, the real facts should eventually come to light. As Shakespeare observed, "The truth will out."[524] So don't be afraid—of flying or other things that go bump in the night. In the words of two-time Nobel Prize winner, Marie Curie,

Nothing in life is to be feared. It is only to be understood.[525]

And during that ongoing process of learning, the following of Albert Einstein is a good thing to keep in mind:

Blind obedience to authority is the greatest enemy of the truth.[526]

And there is one conclusion about this case that still makes more sense than anything in the official version; as the former Prime Minister of Malaysia so aptly said:

Someone is hiding something.[527]

While we continue to search and plea for answers, our hearts go out to the families and friends of each of those two hundred thirty-nine people.

NOTES

Epigraph

1 *Internet Encyclopedia of Philosophy: A Peer-Reviewed Academic Resource,* "Validity and Soundness," accessed July 17, 2014, www.iep.utm.edu/val-snd/.

2 Russ Linden quoting Dwight D. Eisenhower, "Reframe, Broaden a Problem," May 4, 2011, www.governing.com/columns/mgmt-insights/reframe-broaden-problems.html.

Introduction

3 Paul Ausick, "Why a Boeing 777 costs $320 million," March 30, 2014, *USA Today* online, www.usatoday.com/story/money/business/2014/03/30/why-a-boeing-777-costs-320-million-dollars/7063805/.

4 Bill Hutchinson, "Ex-Malaysian prime minister says CIA, Boeing 'hiding' missing airplane," May 19, 2014, *New York Daily News* online, http://www.nydailynews.com/news/world/cia-boeing-hiding-missing-plane-pol-article-1.1798229.

5 Hutchinson, "Ex-Malaysian prime minister says CIA, Boeing 'hiding' missing airplane."

6 Ibid.

7 Ibid.

8 *World Population Review,* 2014, http://worldpopulationreview.com/countries/malaysia-population/.

9   Sam Ro, "The 40 Biggest Economies In The World," June 15, 2013, *Business Insider* online, www.businessinsider.com/largest-economies-world-gdp-2013-6.

10 *Osnet Daily,* "BOMBSHELL: BOEING warned of Cyberjacking, China deploys satellites to search for Flight 370," March 11, 2014, http://osnetdaily.com/2014/03/bombshell-boeing-warned-of-cyberjacking-china-deploys-satellites-to-search-for-flight-370/; Deborah Dupre, "Malaysia jet hidden by Electronic Weaponry? 20 EW defense-linked passengers," March 9, 2014, *Examiner.com*, www.examiner.com/article/malaysia-jet-hidden-by-electronic-weaponry-20-ew-defense-linked-passengers; *Osnet Daily,* "REVEALED: Flight 370 Communication Transcript strengthens Cyberjacking theory," March 22, 2014, http://osnetdaily.com/2014/03/revealed-flight-370-communication-transcript-strengthens-cyberjacking-theory/.

11 Marc Weber Tobias, "The Oddity of Malaysian Airlines Flight 370: Planes Want To Be Seen," April 17, 2014, *Forbes* online, www.forbes.com/sites/marcwebertobias/2014/04/17/the-oddity-of-malaysian-airlines-flight-370-planes-want-to-be-seen/2/.

12 Ibid.

13 Michael Martinez quoting Richard Quest, "Key moments emerge in tracking of missing Malaysia Airlines plane," March 23, 2014, *CNN World* online, http://www.cnn.com/2014/03/15/world/asia/malaysia-airlines-flight-370-chronology/.

14 Tobias, "The Oddity of Malaysian Airlines Flight 370: Planes Want To Be Seen."

15 Ibid.

16 Ibid.

17 Walter Lippmann, *Public Opinion* (Harcourt, Brace and Company, 1922).

18 Edward S. Herman and Noam Chomsky, *Manufacturing Consent: The Political Economy of the Mass Media* (Random House, 1988).

19 Ibid.

20 Noam Chomsky, *Media Control: The Spectacular Achievements of Propaganda* (Open Media, 2003).

21 Richard Belzer and David Wayne, *Dead Wrong: Straight Facts on the Country's Most Controversial Cover-Ups* (Skyhorse, 2012), 110–125.

22 Walt Brown, PhD, *The Warren Omission*, (Delmax, 1996).

23 Belzer and Wayne, *Dead Wrong: Straight Facts on the Country's Most Controversial Cover-Ups*.

24 Richard Belzer, *UFOs, JFK, and Elvis: Conspiracies You Don't Have To Be Crazy To Believe* (Random House, 1999), 196.

25 Hutchinson, "Ex-Malaysian prime minister says CIA, Boeing 'hiding' missing airplane."

26 Dean Garrison, "Philip Wood's Girlfriend Believes Flight 370 is Intact and Passengers are Alive," April 5, 2014 (citing "Girlfriend Of Passenger On Flight 370 Says A Family Member Saying Plane Was Followed By Fighter Jets"), CNN online, www.dcclothesline.com/2014/04/05/philip-woods-girlfriend-believes-flight-370-tact-passengers-alive-2/.

27 *Boeing.com*, "777 Family: Boeing 777 Facts," retrieved August 16, 2014, www.boeing.com/boeing/commercial/777family/pf/pf_facts.page.

One: The Boeing 777-200ER

28 Ibid.

29 Ibid.

30 Gregg F. Bartley, *The Avionics Handbook, Boeing B-777: Fly-By-Wire Flight Controls* (CRC Press, 2001), www.davi.ws/avionics/TheAvionicsHandbook_Cap_11.pdf.

31 *Aviation beta*, "How does IDENT work?" retrieved August 19, 2014, http://aviation.stackexchange.com/questions/3049/how-does-ident-work.

32 John P. Choisser, *MH370: Lost in the Dark: In Defense of the Pilots: An Engineer's Perspective* (JohnChoisser.com, 2014).

33 Patrick Smith, "The Mystery of Malaysia Airlines Flight 370," March 30, 2014, Ask the Pilot, www.askthepilot.com/malaysia-airlines-flight-370/.

34 Patrick Smith, "The Mystery of Malaysia Airlines Flight 370," May 7, 2014, (emphasis in original) Ask the Pilot, www.askthepilot.com/malaysia-airlines-flight-370/.

35 Tobias, "The Oddity of Malaysian Airlines Flight 370: Planes Want To Be Seen."

36 David Soucie and Ozzie Cheek, *Why Planes Crash: An Accident Investigator Fights for Safe Skies* (Skyhorse, 2011).

37 http://transcripts.cnn.com/TRANSCRIPTS/1403/11/cnr.05.html.

38 *CBS News*, "Why can plane transponders be turned off from the cockpit?," March 19, 2014, www.cbsnews. com/news/malaysia-airlines-flight-370-mystery-raises-new-transponder-questions/.

39 Choisser, *MH370: Lost in the Dark: In Defense of the Pilots: An Engineer's Perspective.*

40 Patrick Smith, "The Mystery of Malaysia Airlines Flight 370," April 4, 2014, (emphasis in original) Ask the Pilot, www.askthepilot.com/malaysia-airlines-flight-370/.

41 Choisser, *MH370: Lost in the Dark: In Defense of the Pilots: An Engineer's Perspective.*

42 Ibid.

43 Ibid.

44 Ibid.

45 Ibid.

46 Ibid.

47 Ibid.

48 Ibid.

49 Ibid.

50 Ibid.

51 Ibid.

52 Ibid.

53 Ibid.

54 John Goglia, "Malaysian Air Flight MH370: How Can A Boeing-777 Aircraft Suddenly Lose All Contact?" March 8, 2014, *Forbes* online, www.forbes.com/sites/johngoglia/2014/03/08/malaysian-air-flight-mh370-how-can-a-boeing-777-aircraft-suddenly-lose-all-contact/.

55 Boeing.com, "777 Family: Boeing 777 Facts."

56 Choisser, *MH370: Lost in the Dark: In Defense of the Pilots: An Engineer's Perspective.*

57 Smith, *Cockpit Confidential: Everything You Need To Know About Air Travel: Questions, Answers, and Reflections*, 122.

58 Choisser, *MH370: Lost in the Dark: In Defense of the Pilots: An Engineer's Perspective.*

59 Ibid.

60 Ibid.

61 Ibid.

62 Ibid.

63 Choisser, *MH370: Lost in the Dark: In Defense of the Pilots: An Engineer's Perspective.*

64 Ibid.

65 Matt Wuillemin, "B777 E/E ACCESS – MSc. RESEARCH (Copyright – M WUILLEMIN," June 11, 2012, YouTube, www.youtube.com/watch?v=mLmzvF2qkDY.

66 Ibid.

67 Ibid.

68 Choisser, *MH370: Lost in the Dark: In Defense of the Pilots: An Engineer's Perspective.*

69 Patrick Smith, "What if somebody opens a door during flight?" retrieved August 31, 2014, (emphasis in original) Ask the Pilot, www.askthepilot.com/questionanswers/exits/.

70 Marshall Brain, "What if someone shot a gun on an airplane," retrieved August 31, 2014, HowStuffWorks, http://science.howstuffworks.com/innovation/science-questions/gun-on-plane.htm.

71 Deane Barker, "The Truth About Explosive Decompression," July 7, 2004, Gadgetopia, http://gadgetopia.com/post/2606#sthash.WexM0How.dpuf.

72 Ibid.

73 Ibid.

74 Goglia, "Malaysian Air Flight MH370: How Can A Boeing-777 Aircraft Suddenly Lose All Contact?"

Two: Malaysia Airlines

75 Siva Govindasamy and Alwyn Scott, "Malaysia Airlines has one of Asia's best safety records," March 8, 2014, *Reuters*, www.reuters.com/article/2014/03/08/us-malaysia-airlines-profile-idUSBREA2707Y20140308.

76 Tony Dawber, "Missing Malaysia Airlines flight: Safety record of MAS marred by a SERIES of major incidents," March 12, 2014, *Mirror* online, www.mirror.co.uk/news/world-news/missing-malaysia-airlines-flight-safety-3232586#.U-ROd_ldUuE.

77 Ibid.

78 Ibid.

79 Ibid.

Three: Events of March 8, 2014

80 Ibid.

81 Boeing.com, "777 Family: Boeing 777 Facts."

82 *The Straits Times* online (Singapore), "Vietnam says it told Malaysia that missing plane MH370 had turned back," March 12, 2014, www.straitstimes.com/the-big-story/missing-mas-plane/story/vietnam-says-it-told-malaysia-missing-plane-mh370-had-turned-b.

83 CNN, "Transcript: Malaysian Prime Minister's statement on Flight 370," March 15, 2014 (underlined emphasis added), www.cnn.com/2014/03/15/world/asia/transcript-malaysia-prime-minister/.

84 John P. Choisser, *MH370: Lost in the Dark.*

85 J. Kyle O'Donnell and Dave Merrill, "Missing-Jet Mystery: Flight 370 Timeline," 2014, www.bloomberg.com/infographics/2014-03-13/malaysian-air-flight-370-timeline.html.

86 Ann Colwell, "Timeline of Malaysia Airlines Flight 370," March 12, 2014, www.cnn.com/2014/03/11/world/asia/malaysia-airlines-flight-370-timeline/.

Four: "Official Conclusion," The Plane Crashed into the Indian Ocean

87 Dato' Azharuddin Abrul Rahman, Director General, Department of Civil Aviation, "Press Conference: MH370," March 10, 2014, www.mot.gov.my/my/Newsroom/Press%20Release/Tahun%202014/MH370%20Press%20Statement%20by%20Dato%27%20Azharuddin%20Abdul%20Rahman%20on%2010%20March%202014%20(12.00PM).pdf.

88 Rajeev Sharma, "Connecting the dots: Missing Malaysia Airlines plane a terror attack aimed at China?" March 18, 2014, http://rt.com/op-edge/malaysia-plane-terror-attack-550/.

89 Ibid.

90 Ibid.

91 Ibid.

92 Belzer and Wayne, *Dead Wrong: Straight Facts on the Country's Most Controversial Cover-Ups.*

93 *Daily Mail Reporter*, "Missing jet changed course 12 minutes BEFORE co-pilot's calm last message, say US officials," March 19, 2014, www. dailymail.co.uk/news/article-2584030/Missing-jet-changed-course-12-minutes-BEFORE-pilots-calm-message-say-US-officials.html.

94 Ibid.

95 Ibid.

96 Daniel Stacey, Andy Pasztor, and David Winning, "Australian Report Postulates Malaysia Airlines Flight 370 Lost Oxygen," June 26, 2014, *Wall Street Journal* online, http:// online.wsj.com/articles/malaysia-airlines-flight-370-search-shifts-to-new-area-1403762350.

97 *EducatorHQ*, "Hypoxia – Perfect Painless Death," June 20, 2013, www. youtube.com/watch?v=GOouRavcANQ.

98 *Reuters*, "The Hunt for Malaysia Airlines Flight MH370 Continues," *Newsweek* online, August 6, 2014, www. newsweek.com/hunt-malaysia-airlines-flight-mh370-continues-263291.

Five: Multiple Eyewitness Testimonies You Probably Haven't Heard About

99 Andrew Griffin, "Oil rig worker fired after claiming to have seen burning MH370 plane crash," June 8, 2014, *The Independent* online, www.independent.co.uk/news/world/asia/oil-rig-worker-fired-after-claiming-to-have-seen-burning-mh370-plane-crash-9508187.html.

100 Richard Shears and James Rush, "'Tell us the truth': Chinese families hurl water bottles at Malaysian airline staff as clueless officials admit the plane could be ANYWHERE within 27,000 square nautical mile area," March 12, 2014, *Daily Mail* online, www.dailymail.co.uk/news/article-2578914/Nine-fresh-witnesses-place-missing-jet-near-Thailand-despite-Malaysia-military-moving-search-area-west.html.

101 Email from Mike McKay can be viewed at: The Wire Abby Ohlheiser, "Oil Rig Worker Thinks He Saw Malaysia Air Flight 370 Go Down

in Flames," March 12, 2014, *The Wire* online, www.thewire.com/ global/2014/03/oil-rig-worker-says-he-saw-malaysia-air-flight-370-go-down/359093/.

102 Graeme-Lion, "Oil rig worker…What happened to him," *Reddit: MH 370*, retrieved August 10, 2014, www.reddit.com/r/MH370/ comments/23brbu/oil_rig_workerwhat_happened_to_him/.

103 Sy Gunson, "Oil rig worker…What happened to him," *Reddit: MH 370*, retrieved August 10, 2014: www.reddit.com/r/MH370/ comments/23brbu/oil_rig_workerwhat_happened_to_him/.

104 Alberto Riva, "Email From Worker Saying He Saw Malaysian 370 Go Down Only Adds To Mystery," March 12, 2014, *International Business News* online, www.ibtimes.com/email-worker-saying-he-saw-malaysian-370-go-down-only-adds-mystery-1561068.

105 Choisser, *MH370: Lost in the Dark: In Defense of the Pilots: An Engineer's Perspective.*

106 Riva, "Email From Worker Saying He Saw Malaysian 370 Go Down Only Adds To Mystery."

107 Michael Martinez, "Flying low? Burning object? Ground witnesses claim they saw Flight 370," March 20, 2014, CNN online, www.cnn.com/2014/03/19/world/asia/ malaysia-airlines-plane-ground-witnesses/.

108 Ibid.

109 Shears and Rush, "'Tell us the truth': Chinese families hurl water bottles at Malaysian airline staff as clueless officials admit the plane could be ANYWHERE within 27,000 square nautical mile area."

110 Shears and Rush, "'Tell us the truth': Chinese families hurl water bottles at Malaysian airline staff as clueless officials admit the plane could be ANYWHERE within 27,000 square nautical mile area."

111 *Bernama*, "Missing MH370: Terengganu police receive report on explosion in Marang," March 12, 2014, *New Straits Times* online, www2. nst.com.my/latest/font-color-red-missing-mh370-font-terengganu-police-receive-report-on-explosion-in-marang-1.509347.

112 Alberto Riva, "Malaysia Airlines Flight Spotted In Maldives? Examining The Latest Theory On MH370," March 18, 2014, *International Business Times* online, www.ibtimes.com/malaysia-airlines-flight-spotted-maldives-examining-latest-theory-mh370-1562221.

113 Dean Garrison, "Media Cover Up? Jumbo Jet Witnessed Heading Toward Diego Garcia on March 8," April 10, 2014, *D.C. Clothesline*, www.dcclothesline.com/2014/04/10/media-cover-low-flying-passenger-jet-witnessed-heading-toward-diego-garcia-march-8/.

114 Martinez, "Flying low? Burning object? Ground witnesses claim they saw Flight 370."

115 Ibid.

116 Martinez, "Flying low? Burning object? Ground witnesses claim they saw Flight 370," quoting former Malaysian transport minister Hishammuddin Hussein.

117 William Shakespeare, *Hamlet* (1602), http://en.wikipedia.org/wiki/The_lady_doth_protest_too_much,_methinks.

118 Garrison, "Media Cover Up? Jumbo Jet Witnessed Heading Toward Diego Garcia on March 8."

119 Riva, "Malaysia Airlines Flight Spotted In Maldives? Examining The Latest Theory On MH370."

Six: Media Coverage: You Have Got to Be Kidding

120 Neil Lawrence, "The Death of Newspapers," *Midwest Today* online, May-June-July 2014, 4-9, www.flippubs.com/publication/?i=209054.

121 Joe Coscarelli, "Real CNN Poll Asks: Did 'Space Aliens, Time Travelers or Beings From Another Dimension' Make Flight 370,'" May 7, 2014, http://nymag.com/daily/intelligencer/2014/05/cnn-poll-did-aliens-make-flight-370-disappear.html.

122 Ibid.

123 Ibid.

124 Ibid.

125 Adam K. Raymond, "CNN's 9 Most Deplorable Malaysia Airlines Flight 370 Moments," April 15, 2014, http://nymag.com/daily/intelligencer/2014/04/cnns-9-most-deplorable-flight-370-moments.html.

126 Jon Stewart, "No News Bearers," April 1, 2014, *The Daily Show* online, http://thedailyshow.cc.com/videos/bm3dgu/no-news-bearers.

127 Jon Stewart, "The Duh Room," April 1, 2014, *The Daily Show*: http://thedailyshow.cc.com/videos/f1swql/no-news-bearers---the-duh-room.

128 Jason Easley, "Bill Maher Obliterates CNN's Ghoulish Continuing Coverage of Malaysia Flight 370," May 3, 2014, www.politicususa.com/2014/05/03/bill-maher-obliterates-cnns-goulish-continuing-coverage-malaysia-flight-370.html.

129 Matt Yoder," The Final Days of CNN's Malaysia Airlines Obsession," http://bloguin.com/theapparty/2014-articles/the-final-days-of-cnns-malaysia-airlines-obsession.html, Bloguin.

130 Alexandra Dipalma, "The Best (Worst) Moments of CNN's Malaysia Airlines Flight 370 Coverage, April 18, 2014, http://fusion.net/culture/story/best-worst-moments-cnns-malaysia-airlines-flight-370-604982.

131 Blaire, "CNN's 9 Most Deplorable Malaysia Airlines Flight 370 Moments."

132 *CBS News* online, "Why can plane transponders be turned off from the cockpit?" March 19, 2014, www.cbsnews.com/news/malaysia-airlines-flight-370-mystery-raises-new-transponder-questions/.

133 Martinez, "Key moments emerge in tracking of missing Malaysia Airlines plane."

134 Marc Weber Tobias, "The Oddity of Malaysian Airlines Flight 370: Planes Want To Be Seen," April 17, 2014, *Forbes,* www.forbes.com/sites/marcwebertobias/2014/04/17/the-oddity-of-malaysian-airlines-flight-370-planes-want-to-be-seen/2/.

135 Choisser, *MH370: Lost in the Dark: In Defense of the Pilots: An Engineer's Perspective.*

136 *Turbulence.*

137 Marnie Hunter, "Flight attendant downplays role in helping land airliner," June 16, 2010, *CNN:* www.cnn.com/2010/TRAVEL/06/16/flight.attendant.landing/.

138 *Wikipedia,* "Talk down aircraft landing," August 31, 2014, http://en.wikipedia.org/wiki/Talk_down_aircraft_landing.

139 Smith, *Cockpit Confidential: Everything You Need To Know About Air Travel: Questions, Answers, and Reflections.*

140 Boeing.com, "777 Family: Flight Deck and Airplane Systems," retrieved August 19, 2014, www.boeing.com/boeing/commercial/777family/background/back6.page.

141 *Turbulence.*

142 CNN online, various articles.

143 Mike M. Ahlers, "Why didn't Flight 370's emergency beacon work? Lack of signal buoys hope," April 25, 2014, CNN online, http://www.cnn.com/2014/04/25/world/asia/malaysia-airlines-flight-370-beacons/index.html.

Seven: Public Reaction: Maybe There Is Hope After All

144 Lincoln Feast, "Malaysia jet passengers likely suffocated, Australia says," June 26, 2014, Reuters online, http://news.yahoo.com/malaysia-jet-passengers-likely-suffocated-australia-says-001915883—finance.html.

145 Various Yahoo! users comments on "Malaysia jet passengers likely suffocated, Australia says," Lincoln Feast, June 26, 2014, Reuters online.

146 *Reuters* online, "Comments," July 2, 2014, www.reuters.com/article/comments/idUSKBN0F10FE20140627.

147 Various Ask the Pilot users, "COMMENTS, The Mystery of Malaysia Airlines Flight 370," retrieved August 21, 2014, www.askthepilot.com/malaysia-airlines-flight-370/.

148 Charles Weinacker, University of South Alabama (as comment to: "Malaysia Airlines Flight 370: What People Have Got Wrong"), April 8, 2014, www.flyingmag.com/blogs/going-direct/malaysia-airlines-flight-370-what-people-have-got-wrong.

149 Kevin McConnell (as response to: "Malaysia Airlines Flight 370: What People Have Got Wrong"), April 8, 2014, www.flyingmag.com/blogs/going-direct/malaysia-airlines-flight-370-what-people-have-got-wrong.

Eight: Anomalies: What's Wrong with This Picture?

150 *Huffington Post* online, "Tsunami Debris," retrieved September 17, 2014, www.huffingtonpost.com/tag/tsunami-debris/.

151 Michelle Crouch quoting "Pilot, South Carolina," "13+ Things Your Pilot Won't Tell You," *Reader's Digest*, September 13, 2014, www.rd.com/advice/work-career/50-secrets-your-pilot-wont-tell-you/2/.

152 *Transportation Safety Board of Canada*, "Aviation Investigation Report A98H0003," modified July 27, 2012, www.tsb.gc.ca/eng/rapports-reports/aviation/1998/a98h0003/01report/01factual/rep1_12_01.asp.

153 Ibid.

154 Airliners.net, "The Wings of the Web, The McDonnell-Douglas MD-11," retrieved September 13, 2014, www.airliners.net/aircraft-data/stats.main?id=112.

155 *Aviation Safety Network*, "ATC transcript Swissair Flight 111 – 02 SEP 1998," October 16, 2004, http://aviation-safety.net/investigation/cvr/transcripts/atc_sr111.php.

156 *Code 7700*, "Abnormal Procedures: Declaring an Emergency," retrieved September 18, 2014, http://code7700.com/emergency.html.

157 Carmen Fishwick, "What happened to MH370? A pilot and a flight attendant give their views," March 21, 2014, *The Guardian* online, www.theguardian.com/world/2014/mar/21/what-happened-to-flight-mh370-missing-plane.

158 Ibid.

159 Choisser, *MH370: Lost in the Dark: In Defense of the Pilots: An Engineer's Perspective.*

160 *Flight 370: The Missing Links*, TV Movie, ITN Productions-Discovery Communications.

161 Simon Gunson, private message to author, September 5, 2014.

162 Garrison, "Philip Wood's Girlfriend Believes Flight 370 is Intact and Passengers are Alive"

163 *Famagusta Gazette*, "Helios Crash: Background information," retrieved September 4, 2014, http://famagusta-gazette.com/helios-crash-background-information-p237-69.htm.

164 Ibid.

165 Dean Garrison, "Philip Wood's Girlfriend Believes Flight 370 is Intact and Passengers are Alive," April 5, 2014, *D.C. Clothesline*, www.dcclothesline.com/2014/04/05/philip-woods-girlfriend-believes-flight-370-tact-passengers-alive-2/.

166 Greg Wood, "Malaysia reveals how long lost jetliner went unnoticed," May 1, 2014, *CBS News*: www.cbsnews.com/news/malaysia-airlines-flight-370-vanishing-went-unnoticed-for-17-minutes/.

167 Ibid.

168 Ibid.

169 Sina.com, "The key of questions from family members," April 16, 2014, http://blog.sina.com.cn/s/blog_12ece77a00101eh9v.html.

170 Simon Tomlinson, "Missing jet WAS carrying highly flammable lithium batteries: CEO of Malaysian Airlines finally admits to dangerous cargo four days after DENYING it," March 21, 2014, *DailyMail*, www.dailymail.co.uk/news/article-2586308/Missing-jet-WAS-carrying-highly-flammable-lithium-batteries-CEO-Malaysian-Airlines-finally-admits-dangerous-cargo.html.

171 *The Times of India*, "Mystery of lithium ion batteries in flight MH370 disappearance," May 3, 2014, http://timesofindia.indiatimes.com/world/rest-of-world/Mystery-of-lithium-ion-batteries-in-flight-MH370-disappearance/articleshow/34577514.cms.

172 Athena Yenko, "MH370: Motorola Owns 200kg Lithium Ion Batteries, Source Claims," May 5, 2014, *International Business Times*: http://au.ibtimes.com/articles/550999/20140505/mh370-200kg-batts-motorola-preliminary-report.htm#.VAjWhvldUuF.

173 Ibid.

174 Lindsay Murdoch, "Police investigate possible poisoning of food on missing plane," April 3, 2014, *The Sydney Morning Herald,*" www.smh.com.au/world/police-investigate-possible-poisoning-of-food-on-missing-plane-20140403-zqq33.html.

175 Michael Forsythe and Michael S. Schmidt, "Radar Suggests Jet Shifted Path More Than Once," March 14, 2014, *The New York Times*; *Daily Mail Reporter*, "Missing jet changed course 12 minutes BEFORE co-pilot's calm last message, say US officials," March 19, 2014: www.dailymail.co.uk/news/article-2584030/Missing-jet-changed-course-12-minutes-BEFORE-pilots-calm-message-say-US-officials.html.

176 *Malaysia Chronicle*, "CONFIRMED: M'sian radar was wrong about MH370 – plane didn't do 'kamikaze' dive as claimed," June 24, 2014: www.malaysiachronicle.com/index.php?option=com_k2and-view=itemandid=308042:confirmed-malaysian-radar-was-wrong

-about-mh370-plane-did-not-do-kamikaze-diveandItemid=2#ax-zz3DhldOPoa.

177 Simon Gunson, private message to author, September 5, 2014.

178 Fishwick, "What happened to MH370? A pilot and a flight attendant give their views".

179 Rupa Haria and Jeremy Torr, "MH370 Deliberately Diverted–Malaysian PM," March 15, 2014, *Aviation Week*: http://aviationweek.com/commercial-aviation/mh370-deliberately-diverted-malaysian-pm.

180 Inmarsat, "Inmarsat Statement on Malaysia Airlines flight MH370," March 14, 2014, www.inmarsat.com/news/inmarsat-statement-malaysia-airlines-flight-mh370/.

181 Chris Buckley and Nicola Clark, "Satellite Firm Says Its Data Could Offer Location of Missing Flight," March 14, 2014, *The New York Times*: www.nytimes.com/2014/03/15/world/asia/missing-malaysia-airlines-flight-370.html.

182 Heather Saul, "Missing Malaysia Airlines Flight MH370: Military radar shows jet could have been 'hijacked' and then flown towards Andaman Islands," March 14, 2014, *The Independent*: www.independent.co.uk/news/world/asia/missing-malaysia-airlines-flight-mh370-sources-claim-military-radar-data-shows-plane-was-diverted-to-andaman-islands-9192974.html.

183 *Malaysia Chronicle*, "CONFIRMED: M'sian radar was wrong about MH370—plane didn't do 'kamikaze' dive as claimed".

184 Simon Gunson, "Evidence of a Massive Hoax by the Malaysian Government," revised October 26, 2014, originally a comment at: www.theguardian.com/world/2014/oct/04/mh370-plan-for-search-to-restart-on-sunday-after-four-months-on-hold#start-of-comments.

185 Wood, quoting Australian Air Chief Marshal Angus Houston, "Malaysia reveals how long lost jetliner went unnoticed."

186 *BBC News Asia*, "Missing Malaysia plane MH370: What we know," September 8, 2014, www.bbc.com/news/world-asia-26503141.

187 Eveline Danubrata and Niluksi Koswanage, "A baffling turn in MH370 mystery: Military radar detects flight far off course from last point of radio contact," March 11, 2014, *Reuters*: http://news.nationalpost.com/2014/03/11/a-baffling-turn-in-mh370-mystery-radar-detects-flight-far-off-course-from-last-point-of-radio-contact/.

188 Ibid.

189 *BBC News Asia*, "Missing Malaysia plane MH370: What we know".

190 Danubrata and Koswanage, "A baffling turn in MH370 mystery: Military radar detects flight far off course from last point of radio contact".

191 Ibid.

192 *BBC News Asia*, "Missing Malaysia plane MH370: What we know".

193 Simon Gunson, private message to author, September 18, 2014.

194 Richard Quest, "MH370: Is Inmarsat right?," May 27, 2014, *CNN*: www.cnn.com/2014/05/27/world/asia/mh370-is-inmarsat-right-quest-analysis/index.html.

195 *Wikipedia*, "Malaysia Airlines Flight 370," retrieved September 27, 2014, http://en.wikipedia.org/wiki/Malaysia_Airlines_Flight_370.

196 WikiCommons, "Map of search for MH370", retrieved September 27, 2014, http://upload.wikimedia.org/wikipedia/commons/thumb/4/4e/Map_of_search_for_MH370.png/1280px-Map_of_search_for_MH370.png.

197 *Malaysia Chronicle*, "CONFIRMED: M'sian radar was wrong about MH370 – plane didn't do 'kamikaze' dive as claimed".

198 Jeff Wise, "Why Didn't the Missing Airliner's Passengers Phone for Help?," March 17, 2014: www.slate.com/blogs/future_tense/2014/03/17/malaysia_airlines_flight_370_why_didn_t_the_passengers_phone_for_help.html.

199 David Ray Griffin, "Ted Olson's Report of Phone Calls from Barbara Olson On 9/11: Three Official Denials," April 1, 2008, *GlobalResearch*: www.globalresearch.ca/ted-olson-s-report-of-phone-calls-from-barbara-olson-on-9-11-three-official-denials/8514.

200 Ibid.

201 Wise, "Why Didn't the Missing Airliner's Passengers Phone for Help?".

202 Griffin, "Ted Olson's Report of Phone Calls from Barbara Olson on 9/11: Three Official Denials".

203 Wise, "Why Didn't the Missing Airliner's Passengers Phone for Help?".

204 Wise, "Why Didn't the Missing Airliner's Passengers Phone for Help?".

205 Ibid.

206 Ibid.

207 Staff and agencies, "MH370: satellite phone call revealed as Australia gives update on search," August 28, 2014, *theguardian.com*: www.theguardian.com/world/2014/aug/28/mh370-satellite-phone-call-revealed-as-australia-gives-update-on-search.

208 Holly Yan, Steve Almasy and Catherine E. Shoichet, "Malaysia Airlines Flight 370: Co-pilot's cell phone was on, U.S. official says," April 14, 2014, CNN, www.cnn.com/2014/04/14/world/asia/malaysia-airlines-plane/.

209 Ibid.

210 Ibid.

211 Ian Drury and Candace Sutton, "Why did somebody doctor photo of men who took Flight MH370? Fears of a cover-up amid claims of pictures show passengers with the same set of legs," March 23, 2014, *DailyMail*, www.dailymail.co.uk/news/article-2587554/Did-somebody-doctor-photo-men-took-Flight-MH370-Fears-cover-amid-claims-pictures-passengers-set-legs.html#ixzz3CsNlsqGf.

212 Ibid.

213 Ibid.

214 Malay Mail Online, "Cops find five Indian Ocean practice runways in MH370 pilot's simulator, BH reports," March 18, 2014, www.the-malaymailonline.com/malaysia/article/cops-find-five-indian-ocean-practice-runways-in-mh370-pilots-simulator-bh-r.

215 Ibid.

216 Ibid.

217 Fishwick, "What happened to MH370? A pilot and a flight attendant give their views".

218 James Rush, "Agony of the wife whose husband gave her his wedding ring and watch before boarding missing flight MH370 'in case something should happen to him,'" March 11, 2014, *DailyMail*: www.dailymail.co.uk/news/article-2578358/Husband-gave-wedding-ring-watch-wife-boarding-missing-flight-MH370-case-happen-him.html.

219 Malaysia Airlines, "MH 370 Passenger Manifest," April 10, 2014, www.malaysiaairlines.com/content/dam/malaysiaairlines/mas/PDF/MH370/Malaysia%20Airlines%20Flight%20MH%20370%20Passenger%20Manifest_Nationality@10Apr.pdf.

220 Jim Stone, "Don't believe the black box ping story," April 8, 2014, *Jim Stone- Freelance Journalist*: http://jimstonefreelance.com/phillip-wood.html.

221 Dean Garrison, "Flight 370 Passenger Philip Wood Allegedly Sends Message from U.S. Military Base in Indian Ocean," March 31, 2014, *D.C. Clothesline*: www.dcclothesline.com/2014/03/31/flight-370-passenger-philip-wood-allegedly-sends-message-u-s-military-base-indian-ocean/.

222 Jim Stone, "Remember the shills," April 7, 2014, *Jim Stone- Freelance Journalist*: http://jimstonefreelance.com/phillipwood.html.

223 Henry Austin, "Flight 370: Philip Wood's Girlfriend Sara Bajc Got Death Threat," May 8, 2014, *NBC News*: www.nbcnews.com/storyline/missing-jet/flight-370-philip-woods-girlfriend-sarah-bajc-got-death-threat-n100446.

224 Garrison, "Philip Wood's Girlfriend Believes Flight 370 is Intact and Passengers are Alive".

225 Ronan Farrow interview, "Sarah Bajc, partner of Flight 370 passenger Philip Wood," March 28, 2014, MSNBC: www.msnbc.com/ronan-farrow/watch/plane-passengers-partner-still-skeptical-209483331906.

226 Ashley Collman, "'Something is being covered up': Investigation to find missing MH370 has been sabotaged, says American passenger's girlfriend," September 8, 2014, *DailyMail*: www.dailymail.co.uk/news/article-2747936/Something-covered-Investigation-missing-MH370-sabotaged-says-American-passenger-s-girlfriend.html#ixzz3CmbupqOt.

227 Ibid.

228 Ibid.

229 Shawn Helton, "FLIGHT CONTROL: Boeing's 'Uninterruptible Autopilot System', Drones and Remote Hijacking," August 7, 2014, *21ˢᵗ Century Wire*, emphasis in original, http://21stcenturywire.com/2014/08/07/flight-control-boeings-uninterruptible-autopilot-system-drones-remote-hijacking/.

230 Hutchinson, "Ex-Malaysian prime minister says CIA, Boeing 'hiding' missing airplane."

231 Ibid.

232 Anthony Bond, "Flight MH370: Police chief says 'I know what happened to missing Malaysian Airlines plane,'" September 15, 2014, *Mirror online*: www.mirror.co.uk/news/world-news/flight-mh370-police-chief-says-4261253?ICID=FB_mirror_main.

233 Ibid.

234 Ibid.

Nine: Comparisons Between Malaysia Flight 370 and Other Recent Crashes

235 *The Disappearance of Flight MH370: 14 Days That Gripped the World*, Documentary, Produced by Channel 5 (U.K.), March 21, 2014, www.youtube.com/watch?v=Hqkq1F10i4U.

236 *Malaysia 370: The Plane That Vanished*, Documentary, Produced by Smithsonian Channel, 2014, www.youtube.com/watch?v=23Tvgme62pM.

237 Faith Karimi and Mariano Castillo, "Nine aviation mysteries highlight long history of plane disappearances," March 13, 2014, CNN: www.cnn.com/2014/03/13/world/aviation-mysteries/.

238 Christine Negroni, "Wreckage of Air France Jet Is Found, Brazil Says," June 2, 2009, *The New York Times*: www.nytimes.com/2009/06/03/world/europe/03plane.html?_r=0.

239 Ibid.

240 Ibid.

241 *BBC News*, "Bodies from missing plane found," June 6, 2009, http://news.bbc.co.uk/2/hi/americas/8087303.stm.

242 Daniel Michaels and Max Colchester, "Air France Jet Hit Water Largely Intact, Investigators Say," July 3, 2009, *The Wall Street Journal*, http://online.wsj.com/articles/SB124654219866085907.

243 *Bureau* d'Enquêtes et d'Analyses pour la Sécurité de l'Aviation Civile (BEA)(Bureau of Enquiry and Analysis for Civil Aviation Safety), "Flight AF 447 on 1st June 2009," retrieved October 3, 2014, www.bea.aero/en/enquetes/flight.af.447/flight.af.447.php and Inta Instituto Nacional De Tecnica Aerospacial (National Aerospace Institute), "Data Link Messages Hold Clues to Air France Crash," 2009, www.inta.es/noticiaintaenprensa.aspx?Id=82949.

244 Transportation Safety Board of Canada, "Location of debris field," September 20, 2002, http://bst-tsb. gc.ca/eng/medias-media/urgence-emergency/fond-background/ldf.asp.

245 Ibid.

246 Transportation Safety Board of Canada, "Aviation Investigation Report: In-Flight Fire Leading to Collision with Water," September 2, 1998, www.tsb.gc.ca/eng/rapports-reports/aviation/1998/a98h0003/a98h0003.pdf.

247 Transportation Safety Board of Canada, "Aviation Investigation Report A98H0003," modified July 27, 2012, www.tsb.gc.ca/eng/rapports-reports/aviation/1998/a98h0003/01report/01factual/rep1_01_00.asp.

248 Ibid.

249 Transportation Safety Board of Canada, "Aviation Investigation Report A98H0003," September 2, 1998, www.tsb.gc.ca/eng/rapports-reports/aviation/1998/a98h0003/a98h0003.asp.

250 Dylan Stableford, "Wreckage, remains from missing Air Algerie flight AH5017 reported found in Mali," July 24, 2014, *Yahoo! News*, http://news.yahoo.com/air-algerie-ah5017-plane-disappears-115228417.html.

251 *The Associated Press*, "Air Algerie Flight AH5017 crash: Plane 'disintegrated,' French officials say," July 25, 2014, *CBCnews*, www.cbc.ca/news/world/air-algerie-flight-ah5017-crash-plane-disintegrated-french-officials-say-1.2717718.

252 *CIMSS Satellite Blog*, "Air Algerie Plane Crash in Mali," July 24, 2014, University of Wisconsin-Madison, Space Science and Engineering Center, http://cimss.ssec.wisc.edu/goes/blog/archives/16262.

253 Andy Pasztor, Jon Ostrower, Robert Wall and Jason Ng, "Black Boxes Short on Data," July 28, 2014, *The Wall Street Journal*.

254 Olivier Knox, "US tracked missile that brought down Malaysian Airlines Flight 17," July 22, 2014, *Yahoo! News*, http://news.yahoo.com/us-tracked-missile-that-brought-down-malaysian-airlines-flight-17-222751385.html.

255 David Martin, "What U.S. intelligence knows about Flight 17 shoot down," July 22, 2014, *CBS News*, www.cbsnews.com/news/malaysia-airlines-flight-17-what-u-s-intelligence-knows-about-plane-shoot down/.

256 Ibid.

257 Phil Stewart and Mark Hosenball, "U.S. scrambles to determine who fired Russian-made missile at jet," July 19, 2014, *Reuters*, http://in.reuters.com/article/2014/07/18/ukraine-crisis-airplane-intelligence-idINKBN0FN2QK20140718.

258 Peter Baker, Michael R. Gordon and Mark Mazzetti, "U.S. Sees Evidence of Russian Links to Jet's Downing, July 18, 2014, *The New York Times*, www.nytimes.com/2014/07/19/world/europe/malaysia-airlines-plane-ukraine.html?_r=0.

259 *Associated Press in Kingston*, "Search resumes off Jamaica for 'ghost flight' plane that crashed into sea," September 6, 2014, *theguardian.com*, www.theguardian.com/world/2014/sep/06/search-resumes-off-jamaica-ghost-flight-plane.

260 Brad Knickerbocker, "How hypoxia likely brought down Laurence and Jane Glazer's 'ghost plane'," September 7, 2014, *The Christian Science Monitor*, www.csmonitor.com/USA/2014/0907/How-hypoxia-likely-brought-down-Laurence-and-Jane-Glazer-s-ghost-plane.

261 Ibid.

262 Ibid.

263 Ibid.

264 Thomas Adams, "Laurence and Jane Glazer killed as plane crashes in Jamaica," September 5, 2014, *Rochester Business Journal*, www.rbj.net/article.asp?aID=210489.

265 Ibid.

266 Knickerbocker, "How hypoxia likely brought down Laurence and Jane Glazer's 'ghost plane'".

267 Ibid.

268 Adams, "Laurence and Jane Glazer killed as plane crashes in Jamaica".

269 Ibid.

270 Knickerbocker, "How hypoxia likely brought down Laurence and Jane Glazer's 'ghost plane'".

271 Air Accident Investigation and Aviation Safety Board (AAIASB), "Aircraft Accident Report: Helios Airways Flight HCY522,"

November, 2006, http://cfapp.icao.int/fsix/sr/reports/05002960_final_report.pdf.

272 Ibid.

273 Ibid.

274 *SKYbrary*, "Emergency Depressurization: Guidance for Flight Crews," retrieved October 6, 2014, www.skybrary.aero/index.php/Emergency_Depressurisation:_Guidance_for_Flight_Crews.

275 AAIASB, "Aircraft Accident Report: Helios Airways Flight HCY522".

276 Ibid.

277 Ibid.

278 Ibid.

279 Ibid.

280 National Transportation Safety Board, "Loss of Thrust in Both Engines After Encountering a Flock of Birds and Subsequent Ditching on the Hudson River," January 15, 2009, www.ntsb.gov/doclib/reports/2010/AAR1003.pdf.

281 Matthew L. Wald and Al Baker, "1549 to Tower: 'We're Gonna End Up in the Hudson'," January 17, 2009, *The New York Times,* www.nytimes.com/2009/01/18/nyregion/18plane.html?_r=1.

282 Michael J. Sniffen, "Source: Pilot rejected 2 airport landings," January 16, 2009, *San Francisco Chronicle*, http://web.archive.org/web/20090117040946/http://www.sfgate.com/cgibin/article.cgi?f=/n/a/2009/01/16/national/w110551S61.DTL.

283 Wald and Baker, "1549 to Tower: 'We're Gonna End Up in the Hudson'".

284 Mike Brooks, Jeanne Meserve, and Mike Ahlers, " "Airplane crash-lands into Hudson River; all aboard reported safe," January 15, 2009, CNN, www.cnn.com/2009/US/01/15/new.york.plane.crash/index.html.

285 Patrick Smith, *Cockpit Confidential: Everything You Need To Know About Air Travel: Questions, Answers, and Reflections.*

286 Ibid.

287 Ibid.

288 Linda Shiner, "Sully's Tale," February 18, 2009, *AirSpaceMag*, www.airspace-mag.com/as-interview/aamps-interview-sullys-tale-53584029/?no-ist=.

289 *Malaysia 370: The Plane That Vanished*, Documentary, Produced by Smithsonian Channel, 2014, www.youtube.com/watch?v=23Tvgme62pM.

290 Brooks, Meserve, and Ahlers, "Airplane crash-lands into Hudson River; all aboard reported safe".

291 AirDisaster.com, "Special Report: Ethiopian Airlines Flight 961," retrieved October 16, 2014, www.webcitation.org/6AtdKPxWo

292 AirCrashVideo, "Ethiopian Airlines Boeing 767-260ER crash footage," June 14, 2011, www.youtube.com/watch?v=AvtYtvd5x60.

293 *Air Crash Investigations*, "African Hijack- Ethiopian Flight 961," Documentary, Directed by Karl Jason, Cineflix, Season 3, Episode 10, www.youtube.com/watch?v=GDI2Ziy0Gms.

294 *CedarJet201*, "Ethiopian Airlines Boeing 767 hijacker ON Radar + Air Traffic Recordings," February 17, 2014, www.youtube.com/watch?v=k5gnwts6IMU.

Ten: Explanation #1—Weather

295 ABC.net, "Malaysia Airlines MH370: What we know about the missing plane," April 27, 2014, www.abc.net.au/news/2014-03-10/malaysia-airlines-flight-mh370-what-we-know/5309688.

Eleven: Explanation #2—Hijacking

296 ABC.net, "Malaysia Airlines MH370: What we know about the missing plane," April 27, 2014, www.abc.net.au/news/2014-03-10/malaysia-airlines-flight-mh370-what-we-know/5309688.

297 Alberto Riva, "Three Hijackings That Look Like The Case Of Missing Malaysia Airlines 370," March 16, 2014 *International Business Times*, www.ibtimes.com/three-hijackings-look-case-missing-malaysia-airlines-370-1561732.

298 Wikimedia Commons, "File: Malaysia Airlines Flight MH370 cockpit transcript (official 1 April 2014)," April 1, 2014, http://commons.wikimedia.org/w/index.php?title=File:Malaysia_Airlines_Flight_MH370_cockpit_transcript_(official_1_April_2014).pdfandpage=2.

299 Thom Patterson and Catherine E. Shoichet, "What happened to flight 370? Four scenarios fuel speculation among experts," March

10, 2014, CNN, www.cnn.com/2014/03/10/world/asia/malaysia-plane-scenarios/index.html.

300 Ibid. CNN quoting three pilots.

301 Ibid. CNN quoting three pilots.

302 Danubrata and Koswanage, "A baffling turn in MH370 mystery: Military radar detects flight far off course from last point of radio contact," March 11, 2014, *Reuters*, http://news.nationalpost.com/2014/03/11/a-baffling-turn-in-mh370-mystery-radar-detects-flight-far-off-course-from-last-point-of-radio-contact/.

303 Danubrata and Koswanage, "A baffling turn in MH370 mystery: Military radar detects flight far off course from last point of radio contact".

304 *BBC News*, "Missing Malaysia plane: The passengers on board MH370," March 25, 2014, www.bbc.com/news/world-asia-26503469.

305 ABC.net, "Malaysia Airlines MH370: What we know about the missing plane".

306 Ibid.

307 *AP*, "INTERPOL releases image of 2 Iranians on missing jet," March 11, 2014, *The Times of Israel*, www.timesofisrael.com/INTERPOL-releases-image-of-2-iranians-on-missing-jet/.

308 Katia Hetter and Karla Cripps, "Who travels with a stolen passport?" March 11, 2014, CNN, www.cnn.com/2014/03/10/travel/malaysia-airlines-stolen-passports/index.html.

309 Danubrata and Koswanage, "A baffling turn in MH370 mystery: Military radar detects flight far off course from last point of radio contact".

310 Hetter and Cripps, "Who travels with a stolen passport?".

311 Ted Jeory, "Malaysian plane: 20 passengers worked for ELECTRONIC WARFARE and MILITARY RADAR firm," March 19, 2014, *Daily Express*, www.express.co.uk/news/world/465557/Malaysian-plane-20-on-board-worked-for-ELECTRONIC-WARFARE-and-radar-defence-company.

312 Choisser, *Malaysia Flight MH370, Lost in the Dark, In Defense of the Pilots: An Engineer's Perspective*.

313 *Malaysia 370: The Plane That Vanished*, Documentary, Produced by Smithsonian Channel, 2014, www.youtube.com/watch?v=23Tvgme62pM.

Twelve: Explanation #3—Pilot Terrorist Activity, or Suicide

314 Jamie Doward, Kate Hodal and Tania Branigan, "Missing Malaysia Airlines plane 'sabotaged on board.'" March 15, 2014, *The Guardian*, www.theguardian.com/world/2014/mar/15/malaysia-airlines-flight-mh370.

315 Kirk Semple, "Pilots' Possible Role in Flight 370 Vanishing 'Unthinkable' to Friends," March 17, 2014, *The New York Times*, www.nytimes.com/2014/03/18/world/asia/pilots-possible-role-in-flight-370-vanishing-unthinkable-to-friends.html?action=clickandcontentCollection=Asia%20Pacificandmodule=RelatedCoverageandregion=Marginaliaandpgtype=article.

316 Jenni Ryall and Staff Writers, "Malaysian police investigation names MH370 pilot 'prime suspect,'" June 23, 2014, news.com.au, http://mobile.news.com.au/travel/travel-updates/malaysian-police-investigation-names-mh370-pilot-prime-suspect/story-fnizu68q-1226962811653.

317 Paul Thompson, "Recovered Simulator Files Show Possible Devious Intent By MH370 Pilot," June 23, 2014, *Flight Club*, http://flight-club.jalopnik.com/recovered-simulator-files-show-possible-devious-intent-1594670738.

318 Jill Reilly, "FBI analyse Malaysian Airlines pilot's home flight simulator as it's revealed he deleted data one month prior to taking control of missing MH370 plane," March 19, 2014, *DailyMail*, www.dailymail.co.uk/news/article-2584123/Revealed-Malaysian-Airlines-pilot-high-security-US-base-Diego-Garcia-programmed-homemade-flight-simulator-deleted-data-just-taking-control-missing-plane.html#ixzz3Gp7qdCGw.

319 *CBS News/AP*, "Malaysia Airlines Flight 370: Where the investigation stands now," March 27, 2014, www.cbsnews.com/news/malaysia-airlines-flight-370-where-the-investigation-stands-now/.

320 Ibid.

321 news.com.au, "Malaysian police investigation names MH370 pilot 'prime suspect'," June 23, 2014, www.news.com.au/travel/travel-updates/

malaysian-police-investigation-names-mh370-pilot-prime-suspect/
story-fnizu68q-1226962811653.

322 Harriet Alexander, "Missing Malaysia Airlines flight MH370: What
could have happened?," March 18, 2014, *The Telegraph*, www.tel-
egraph.co.uk/news/worldnews/asia/malaysia/10705396/Missing-
Malaysia-Airlines-flight-MH370-What-could-have-happened.html.

323 Simon Parry, Amanda Williams and Wills Robinson, "'Democracy
is dead': 'Fanatical' missing airliner pilot pictured wearing political
slogan T-shirt," March 15, 2014, *DailyMail*, www.dailymail.co.uk/
news/article-2581817/Doomed-airliner-pilot-political-fanatic-
Hours-taking-control-flight-MH370-attended-trial-jailed-
opposition-leader-sodomite.html.

324 Brownie Marie, "Missing Malaysia Flight 370 update: Police
have cleared everyone on flight except the pilot," June 23, 2014,
Christian Today, www.christiantoday.com/article/missing.malaysia.
flight.370.update.police.have.cleared.everyone.on.flight.except.the.
pilot/38354.htm.

325 *Malaysia 370: The Plane That Vanished*, Documentary,
Produced by *Smithsonian Channel*, 2014, www.youtube.com/
watch?v=23Tvgme62pM.

326 Choisser, *MH370: Lost in the Dark: In Defense of the Pilots: An Engineer's
Perspective*.

327 Ibid.

328 *Malaysia 370: The Plane That Vanished*, Documentary, Produced by
Smithsonian Channel.

329 Danubrata and Koswanage, "A baffling turn in MH370 mystery:
Military radar detects flight far off course from last point of radio
contact".

330 *Malaysia 370: The Plane That Vanished*, Documentary, Produced by
Smithsonian Channel.

331 Ibid.

332 Alexander, "Missing Malaysia Airlines flight MH370: What could
have happened?".

333 Marie, "Missing Malaysia Flight 370 update: Police have cleared
everyone on flight except the pilot".

334 Ibid.

335 Ewan Wilson and Geoff Taylor, *Goodnight Malaysian 370: The Truth Behind the Loss of Flight 370*, (Wilson Aviation, Ltd.: 2014).

336 *APNZ*, "MH370 authors hit back at airline," September 23, 2014, *The New Zealand Herald*, www.nzherald.co.nz/nz/news/article.cfm?c_id=1andobjectid=11330078.

337 *The Hans India*, "Mentally Sick MH370 Pilot Committed Suicide: Expert" - .

338 *APNZ*, "MH370 authors hit back at airline".

339 Simon Gunson, private message to author, September 23, 2014, emphasis in original.

340 *Malaysia 370: The Plane That Vanished*, Documentary, Produced by Smithsonian Channel.

341 Choisser, *MH370: Lost in the Dark: In Defense of the Pilots: An Engineer's Perspective*.

342 Malaysian Jet Was in Controlled Flight After Contact Was Lost, Officials Suspect.

343 *Malaysia Chronicle*, "CONFIRMED: M'sian radar was wrong about MH370 – plane didn't do 'kamikaze' dive as claimed"—missing plane likely not seriously damaged over the ocean.

344 *Fox News*, "Here Are the Latest Developments on the Missing Malaysian Plane Part 2," March 14, 2014, www.youtube.com/watch?v=4KS9oaIRi4E.

345 Matthew L. Wald and Michael S. Schmidt, "Lost Jet's Path Seen as Altered via Computer," March 17, 2014, *The New York Times*, www.nytimes.com/2014/03/18/world/asia/malaysia-airlines-flight.html?_r=1.

346 Ibid.

Thirteen: Explanation #4—Mechanical Malfunction/Fire

347 Chris Goodfellow, "A Startlingly Simple Theory About the Missing Malaysia Airlines Jet," March 18, 2014, *Wired*, www.wired.com/2014/03/mh370-electrical-fire/.

348 James Fallows, "Malaysia 370, Day 10: One Fanciful Hypothesis, and Another That Begins to Make Sense," March 17, 2014, *The Atlantic*, www.theatlantic.com/technology/archive/2014/03/malaysia-370-day-10-one-fanciful-hypothesis-and-another-that-begins-to-make-sense/284468/.

349 Alex Davies, "Here's What Pilots Think About The New Idea That The Missing Plane Flew For Hours After A Fire Killed The Pilots," March 18, 2014, *Business Insider*, www.businessinsider.com/did-missing-plane-fly-for-hours-after-deadly-fire-2014-3#ixzz3HBUH-DlmZ.

350 Ibid.

351 Ibid.

352 Ibid.

353 Ibid.

354 *ABC.net*, "Malaysia Airlines MH370: What we know about the missing plane".

355 Jeff Wise, "A 'Startlingly Simple Theory' About the Missing Airliner is Sweeping the Internet. It's Wrong," March 18, 2014, *future tense*, www.slate.com/blogs/future_tense/2014/03/18/mh370_disappearance_chris_goodfellow_s_theory_about_a_fire_and_langkawi.html.

356 *ABC.net*, "Malaysia Airlines MH370: What we know about the missing plane," bold emphasis in original.

357 Wise, "A 'Startlingly Simple Theory' About the Missing Airliner is Sweeping the Internet. It's Wrong," bold emphasis in original.

358 Paul Thompson, "A Problem With The Malaysia Airlines Flight 370 Fire Theory," March 18, 2004, *Flight Club*, http://flightclub.jalopnik.com/a-problem-with-the-malaysia-airlines-flight-370-fire-th-1546500628.

359 Ibid.

360 Ewan Wilson, private message to author, September 24 , 2014.

361 Simon Gunson, private message to author, September 23, 2014.

362 Simon Gunson, private message to author, September 7, 2014.

363 Simon Gunson, "Comments: MH370 underwater search resumes as ships scour Indian Ocean," June 28, 2014, *The Guardian*, https://profile.theguardian.com/user/id/12086801.

364 Choisser, *Malaysia Flight MH370, Lost in the Dark, In Defense of the Pilots: An Engineer's Perspective.*

365 Chris Lee, BSAE, Aerospace Engineer, "What happened to Malaysian Airlines flight 370?" Quora, www.quora.com/What-happened-to-Malaysian-Airlines-flight-MH370.

366 Shawn Helton, "What's Behind the Disappearance of Malaysian Airliner MH370? Part One:," March 11, 2014, *21ˢᵗ Century Wire*, emphasis in original, http://21stcenturywire.com/2014/03/11/whats-behind-the-disappearance-of-malaysian-airliner-mh370/.

367 Ibid, emphasis added.

368 Danubrata and Koswanage, "A baffling turn in MH370 mystery: Military radar detects flight far off course from last point of radio contact".

369 Alexander, "Missing Malaysia Airlines flight MH370: What could have happened?" emphasis in original.

370 Danielle Wiener-Bronner, "A Complete Timeline of the Search for Malaysia Airlines Flight 370," *The Wire*, www.thewire.com/global/2014/03/heres-every-mh370-theory-weve-considered-so-far/359355/.

371 Paul Thompson, "U.N. Nuke Watchdog – No MH370 Explosion Or Plane Crash Detected," March 18, 2014, *Flight Club*, http://flight-club.jalopnik.com/u-n-nuke-watchdog-no-mh370-explosion-or-plane-crash-1546534693.

372 Ibid.

373 Ibid, emphasis added.

Fourteen: Explanation #5—Aliens/Other Extraordinary Means

374 Sebastian Anthony, "We'll find alien life in the next 20 years with our new, awesome telescopes says NASA,' July 15, 2014, *ExtremeTech*, www.extremetech.com/extreme/186321-well-find-alien-life-in-the-next-20-years-with-our-new-awesome-telescopes-says-nasa.

375 Ibid.

376 *U.S. Air Force*, "General Nathan F. Twining," retrieved August 23, 2014, www.af.mil/AboutUs/Biographies/Display/tabid/225/Article/105367/general-nathan-f-twining.aspx, quoting charles bolden.

377 Belzer, *UFOs, JFK, and Elvis: Conspiracies You Don't Have To Be Crazy To Believe.*

378 U.S. Air Force, "General Nathan F. Twining".

379 Nicap.org, "Twining letter," emphasis (bold only) added, (underlining present in original document), retrieved August 23, 2014, www.nicap.org/twining_letter_docs.htm.

380 Belzer, *UFOs, JFK, and Elvis: Conspiracies You Don't Have To Be Crazy To Believe.*

381 Ibid.

382 Charles E. Sellier and Joe Meier, *UFO*, (Contemporary: 1997), 88–89.

383 Ibid, 257.

384 Belzer, *UFOs, JFK, and Elvis: Conspiracies You Don't Have To Be Crazy To Believe.*

385 Clark McClelland, "Senator Barry Goldwater On UFOs, ETs and Roswell," March 23, 2006, www.rense.com/general70/cla.htm.

386 McClelland, "Senator Barry Goldwater On UFOs, ETs and Roswell".

387 Dick Methia, "10 FAMOUS PEOPLE WHO BELEIVE IN UFOS," March 9, 2011, www.listosaur.com/bizarre-stuff/10-famous-people-who-believe-in-ufos/.

388 Methia, "10 FAMOUS PEOPLE WHO BELEIVE IN UFOS"; *The Telegraph*, "Are UFOs real? Famous people who believed," April 22, 2009, www.telegraph.co.uk/technology/5201410/Are-UFOs-real-Famous-people-who-believed.html.

389 Leslie Kean, "Chile Releases Official Study on UFO Photos," July 6, 2014, www.huffingtonpost.com/leslie-kean/government-agency-in-chil_b_5558713.html.

390 Annalee, Newitz, "UFO in Chile Said to be the Size of Two Soccer Stadiums," March 27, 2014, http://io9.com/ufo-in-chile-said-to-be-the-size-of-two-soccer-stadiums-1553302591.

391 Clark C. McClelland, SCO (Space Craft Operator), November 9, 2008, all emphasis in original, www.ufodigest.com/news/1108/tall-et-print.html.

392 McClelland, "Senator Barry Goldwater On UFOs, ETs and Roswell"

393 *MariaMuir.com*, "Cope Tiger: Military War Game Exercise and missing flight MH370," April 17, 2014, http://mariamuir.com/cope-tiger-military-war-game-exercise-and-missing-flight-mh370/.

Fifteen: Explanation #6—Shoot-Down by Missile

394 Matthias Chang, "MH370—A Sinister Tragedy in the Fog of Coincidence? Some strange parallels with catastrophic consequences," April 1, 2014, *FutureFastForward*, http://projectcamelotportal.com/blog/31-kerrys-blog/2076-letter-to-the-world-false-flag-flt-370.

395 *MariaMuir.com*, "Cope Tiger: Military War Game Exercise and missing flight MH370".

396 Ibid.

397 Patterson and Shoichet, "What happened to flight 370? Four scenarios fuel speculation among experts".

398 Michael Shrimpton, "The Shootdown of Malaysian Airlines Flight MH370," March 15, 2014, www.veteranstoday.com/2014/03/15/the-shootdown-of-malaysian-airlines-flight-mh370/comment-page-1/.

399 Nigel Cawthorne, *Flight MH370: The Mystery*, John Blake: 2014.

400 Sam Rkaina, "Malaysia plane crash: Latest disaster reopens questions on whether missing MH370 was shot down," July 17, 2014, *Mirror*, www.mirror.co.uk/news/world-news/malaysia-plane-crash-latest-disaster-3875083#ixzz3FsxtjbkO.

401 Frans Timmermans, Dutch Foreign Minister, "CNN Transcript," October 9, 2014, CNN http://transcripts.cnn.com/TRANSCRIPTS/1410/09/cnr.07.html.

402 CNN, "MH17 victim was wearing oxygen mask," October 9, 2014, http://us.cnn.com/video/data/2.0/video/world/2014/10/09/idesk-intv-soucie-mh17-victim.cnn.html.

403 Ibid.

404 Sophie Shevardnadze, "Lost MH370 flight could've been shot down by missile – aviation expert Col. J. Joseph," March 28, 2014, *RT*, http://rt.com/shows/sophieco/lost-flight-shot down-777/.

405 William Robert Plumlee, private message to author, October 11, 2014.

406 Ibid.

407 Ibid.

408 Ibid.

Sixteen: Explanation #7—Sabotage

409 Deborah Dupre, "Malaysia jet hidden by Electronic Weaponry? 20 EW defense-linked passengers," March 9, 2014, *examiner.com*, http://www.examiner.com/article/malaysia-jet-hidden-by-electronic-weaponry-20-ew-defense-linked-passengers.

410 Jeory, "Malaysian plane: 20 passengers worked for ELECTRONIC WARFARE and MILITARY RADAR firm".

411 Ibid.

412 Ibid.

413 Shepard Ambellas, "Rothschild owned Blackstone Group benefits from missing flight 370, becoming primary patent holder of new technology, reports say," March 25, 2014, *Intellihub*, www.intellihub.com/rothschild-owned-blackstone-group-benefits-missing-flight-370-becoming-primary-patent-holder-new-technology/.

414 Jeory, "Malaysian plane: 20 passengers worked for ELECTRONIC WARFARE and MILITARY RADAR firm".

415 *Snopes.com*, "Patent Pending," March 13, 2014, www.snopes.com/politics/conspiracy/malaysiapatent.asp.

416 Ibid.

417 Ibid.

418 Bill Giovino, "Freescale Semiconductor Bought Out by Blackstone for $17.6 Billion," September 25, 2006, *Microcontroller*, http://microcontroller.com/news/freescale_buyout.asp.

419 Melanie Warner, "What Do George Bush, Arthur Levitt, Jim Baker, Dick Darman, And John Major? Have In Common (They All Work For The Carlyle Group.) What exactly does it do? To find out, we peeked down the rabbit hole." March 18, 2002, *Fortune*, http://archive.fortune.com/magazines/fortune/fortune_archive/2002/03/18/319881/index.htm.

420 William D. Hartung, "Dick Cheney and the Self-Licking Ice Cream Cone, The Carlyle Group: Crony Capitalism without Borders," excerpted from the book, *How Much Are You Making On The War Daddy?: A Quick and Dirty Guide to War Profiteering in the Bush Administration*, Nation Books: 2003, www.thirdworldtraveler.com/Corporate_Welfare/CarlyleGroup_HMOWD%3F.html.

421 Athena Yenko, "MH370: How Fatal is the Chip That Rothschild Reportedly 'Acquired'?" April 14, 2014, *International Business Times*,

http://au.ibtimes.com/articles/547975/20140414/mh370-miss-ing-malaysia-airlines.htm#.VD75gPldXRs.

422 Ibid.

423 Ibid.

424 *Malaysia 370: The Plane That Vanished*, Documentary, Produced by Smithsonian Channel, 2014, www.youtube.com/watch?v=23Tvgme62pM.

Seventeen: Explanation #8—Advanced Technologies

425 David Kerley, Richard Coolidge and Jordyn Phelps, "What would Buck Rogers think? Hi-tech weapons and protection rolled out for Navy," October 6, 2014, *Power Players, ABC News*, http://news.yahoo.com/blogs/power-players-abc-news/technologies-once-avail-able-only-in-movies-are-now-a-reality-for-the-us-navy-225652354.html.

426 Ibid.

427 Ibid.

428 Rob Balsamo, "Interview with Wayne Anderson: Whistleblower Reveals 'Backdoor' 757 Remote Control And Flight Crew 'Lockout' Technology Available Prior To 9/11," retrieved October 6, 2014, http://pilotsfor911truth.org/Remote_Control_Whistleblower.html.

429 *Homeland Security News Wire: John Croft's Flight Global report*, "Boeing wins patent on uninterruptible autopilot system," December 4, 2006, www.homelandsecuritynewswire.com/boeing-wins-patent-uninterruptible-autopilot-system.

430 Ibid.

431 Shawn Helton, "FLIGHT CONTROL: Boeing's 'Uninterruptible Autopilot System', Drones and Remote Hijacking," August 7, 2014, *21ˢᵗ Century Wire*, http://21stcenturywire.com/2014/08/07/flight-control-boeings-uninterruptible-autopilot-system-drones-re-mote-hijacking/.

432 Ibid.

433 Field McConnell and David Hawkins, "Evidence of the Hijack of Malaysian Airways Flight MH370 using the Boeing-Honeywell Uninterruptible Autopilot," July 21, 2014, *Abel Danger*, www.abel-danger.net/2014/07/churchills-red-switch-grandsons-and_4080.html.

434 Christopher Leake, "New autopilot will make another 9/11 impossible," March 3, 2007, *Daily Mail*, www.dailymail.co.uk/news/article-439820/New-autopilot-make-9-11-impossible.html.

435 *Patents*, "System and method for automatically controlling a path of travel of a vehicle," February 19, 2003, www.google.com/patents/US7142971.

436 Ibid.

437 *Patents*, "Method and apparatus for preventing an unauthorized flight of an aircraft," April 16, 2003, www.google.com.au/patents/US7475851.

438 Ibid.

439 *Federal Aviation Administration*, "Special Conditions: Boeing Model 777-200, -300, and -300ER Series Airplanes; Aircraft Electronic System Security Protection From Unauthorized Internal Access," November 18, 2013, www.federalregister.gov/articles/2013/11/18/2013-27343/special-conditions-boeing-model-777-200—300-and—300er-series-airplanes-aircraft-electronic-system.

440 Ibid.

441 *Malaysia 370: The Plane That Vanished*, Documentary, Produced by *Smithsonian Channel*, 2014, www.youtube.com/watch?v=23Tvgme62pM.

442 Wills Robinson, "Is missing Malaysian jet the world's first CYBER HIJACK? Chilling new theory claims hackers could use a mobile phone to take over the controls," March 16, 2014, *DailyMail*, www.dailymail.co.uk/news/article-2582015/Is-missing-Malaysian-plane-world-s-CYBER-HIJACK.html#ixzz3FWXBlBxw.

443 Ibid.

444 Uber Geek, "UK Experts Believe Malaysian Airlines MH370 Was Hijacked Using A Cellphone," March 18, 2014, *Wonderful Engineering*, http://wonderfulengineering.com/uk-experts-believe-malaysian-airlines-mh370-was-hijacked-using-a-cellphone/.

445 Balsamo, "Interview with Wayne Anderson: Whistleblower Reveals 'Backdoor' 757 Remote Control And Flight Crew 'Lockout' Technology Available Prior To 9/11".

446 **Andreas von Bülow**, Die CIA und der 11. September. Internationaler Terror und die Rolle der Geheimdienste, Piper: 2003, www.amazon.

com/September-Internationaler-Terror-Rolle-Geheimdienste/
dp/3492045456.

447 *History Commons*, "Profile: Andreas von *Bülow*," *retrieved October 9, 2014*,
www.historycommons.org/entity.jsp?entity=andreas_von_bulow.

448 NASA.gov, "NASA Armstrong Fact Sheet: Controlled Impact
Demonstration," February 28, 2014, www.nasa.gov/centers/dryden/
news/FactSheets/FS-003-dfrc.html#.VDXUwvldUuF.

449 Ibid.

450 James Corbett, "How to Steal an Airplane: From 9/11 to MH370,"
March 19, 2014, *The Corbett Report*, www.corbettreport.com/
how-to-steal-an-airplane-from-911-to-mh370/.

451 Boeing.com, "Defense, Space and Security: E-4B Advanced Airborne
Command Post," retrieved October 9, 2014, www.boeing.com/boe-
ing/defense-space/military/e4b/index.page.

452 *CNN's Anderson Cooper 360°*, September 12, 2007, CNN, www.you-
tube.com/watch?v=SMK5bmdAEHc.

453 Ibid.

454 Boeing.com, "Defense, Space and Security: E-4B Advanced Airborne
Command Post".

Eighteen: Explanation #9—US Military Air Base, Diego Garcia

455 Cassandra Yeoh, Ron J. Backus and Samuel Lee, "Where
in the world is Diego Garcia," April 9, 2014, www.
thestar.com.my/Lifestyle/Features/2014/04/04/
Where-in-the-world-is-Diego-Garcia/.

456 Ibid.

457 *Princeton University Press*, "Reviews: *Island of Shame: The Secret History
of the U.S. Military Base on Diego Garcia*," retrieved September 26,
2014, http://press.princeton.edu/titles/8885.html.

458 Andy Worthington, "The CIA's Secret Prison on Diego Garcia," August
2–4, 2008, *Counterpunch*, www.counterpunch.org/2008/08/02/
the-cia-s-secret-prison-on-diego-garcia/.

459 *Malaysia 370: The Plane That Vanished*, Documentary, Produced
by Smithsonian Channel, 2014, http://www.youtube.com/
watch?v=23Tvgme62pM.

460 Mike Adams, "Six important facts you're not being told about lost Malaysia Airlines Flight 370," March 10, 2014, bold emphasis in original, www.naturalnews.com/044244_Malaysia_Airlines_Flight_370_vanished.html#ixzz3FxMM3XaG.

461 John Aravosis, "Some questions about Malaysia Air Flight 370," March 13, 2014, bold emphasis in original, http://americablog.com/2014/03/unanswered-questions-malaysia-air-flight-370.html.

462 Tom Allard and Lindsay Murdoch, "The mystery of missing Malaysia Airlines flight MH370," March 19, 2014, *The Sydney Morning Herald*, emphasis added, www.smh.com.au/world/the-mystery-of-missing-malaysia-airlines-flight-mh370-20140314-34sdo.html.

463 Jill Reilly, "FBI analyse Malaysian Airlines pilot's home flight simulator as it's revealed he deleted data one month prior to taking control of missing MH370 plane," March 19, 2014, *Daily Mail*, www.dailymail.co.uk/news/article-2584123/Revealed-Malaysian-Airlines-pilot-high-security-US-base-Diego-Garcia-programmed-homemade-flight-simulator-deleted-data-just-taking-control-missing-plane.html.

464 Ibid.

465 Rachelle Corpuz, "MH370: U.S. Refutes Claims Plane Lands in Diego Garcia," April 12, 2014, *International Business Times*, http://au.ibtimes.com/articles/547834/20140412/malaysia-airlines-flight-mh370-abbott-australia-search.htm#.VDtFh_ldUuE.

466 Shawn Helton, "The Case of Malaysia's Missing Airliner MH370 – Part Two," March 16, 2014, *21st Century Wire*, http://21stcenturywire.com/2014/03/16/the-case-of-malaysias-missing-airliner-mh370-part-two/.

467 Reilly, "FBI analyse Malaysian Airlines pilot's home flight simulator as it's revealed he deleted data one month prior to taking control of missing MH370 plane".

468 Shepard Ambellas, "BREAKING News: Flight 370 Passenger Managed to Send Photo From Hidden iPhone Tracing Back to U.S. Military Base Diego Garcia," March 17, 2014, *Intelihub News*, http://humansarefree.com/2014/04/breking-news-flight-370-passenger.html.

469 *Fox News*, "Gen. McInerney: US hasn't 'come clean' with Flight 370 data," March 19, 2014, http://video.foxnews.com/v/3363882454001/gen-mcinerney-us-hasnt-come-clean-with-flight-370-data/#sp=show-clips.

470 *Fox News*, "Here Are the Latest Developments on the Missing Malaysian Plane Part 2," March 14, 2014, www.youtube.com/watch?v=4KS9oaIRi4E.

471 Ibid.

472 *Fox News*, "Gen. McInerney: US hasn't 'come clean' with Flight 370 data".

473 Ibid.

474 Ibid.

Nineteen: Conclusions

475 *The People's Trust-Malaysia*, "#MH370—SOS distress call was sent from the plane—China Times," May 7, 2014, http://peoplestrustmalaysia.wordpress.com/2014/05/07/mh370-sos-distress-call-was-sent-from-the-plane-china-times/ —citing *China Times*, March 8, 2014, http://www.chinatimes.com/realtimenews/20140308003502-260401.

476 Gunson, citing *New Straits Times*, message to author, September 5, 2014.

477 Shepard Ambellas, "Girlfriend of 370 passenger Wood: 'Fighter jets accompanied flight 370 in secret militarized operation'," April 4, 2014, *Intellihub*, www.intellihub.com/girlfriend-370-passenger-wood-fighter-jets-accompanied-flight-370-secret-militarized-operation-husband-still-alive/ .

478 Martinez, "Flying low? Burning object? Ground witnesses claim they saw Flight 370".

479 Drury & Sutton, "Why did somebody doctor photo of men who took Flight MH370? Fears of a cover-up amid claims of pictures show passengers with the same set of legs".

480 *Malaysia Chronicle*, "CONFIRMED: M'sian radar was wrong about MH370 – plane didn't do 'kamikaze' dive as claimed".

481 Thompson, "U.N. Nuke Watchdog—No MH370 Explosion Or Plane Crash Detected".

482 Wise, "Why Didn't the Missing Airliner's Passengers Phone for Help?".

483 Yan, Almasy & Shoichet, "Malaysia Airlines Flight 370: Co-pilot's cell phone was on, U.S. official says".

484 Jim Stone, "Remember the shills," April 7, 2014, http://jimstonefree-lance.com/phillipwood.html.

485 Yenko, "MH370: How Fatal is the Chip That Rothschild Reportedly 'Acquired'?".

486 Helton, "FLIGHT CONTROL: Boeing's 'Uninterruptible Autopilot System', Drones & Remote Hijacking".

487 *PBS Nova*, "Crash of Flight 447" transcript, February 16, 2011, www.pbs.org/wgbh/nova/space/crash-flight-447.html.

488 Erin Burnett, "MH370 radar data may have been wrong," June 24, 2014, *CNN Erin Burnett OutFront*, www.cnn.com/video/data/2.0/video/bestoftv/2014/06/24/erin-intv-miles-malaysia-mh-370-new-search-area.cnn.html.

489 Danubrata & Koswanage, "A baffling turn in MH370 mystery: Military radar detects flight far off course from last point of radio contact".

490 *ABC.net*, "Malaysia Airlines MH370: What we know about the missing plane".

491 Ibid.

492 Matthew L. Wald & Michael S. Schmidt, "Lost Jet's Path Seen as Altered via Computer," March 17, 2014, *The New York Times*, www.nytimes.com/2014/03/18/world/asia/malaysia-airlines-flight.html?_r=1.

493 *Pure History*, "Malaysia Airlines Flight 370," April 1, 2014, http://purehistory.org/malaysia-airlines-flight-370/ citing "MH370 Press Statement by Ministry of Transport, Malaysia," March 23, 2014, and Wald & Schmidt, "Lost Jet's Path Seen as Altered via Computer": www.mot.gov.my/en/Newsroom/Press%20Release/Year%202014/MH370%20Press%20Statement%20by%20Hishammuddin%20Hussein%20on%2023%20March%202014.pdf.

494 Simon Gunson, private message to author, September 7, 2014.

495 Simon Gunson, private message to author, September 7, 2014.

496 Jeff Wise, "Why Inmarsat's MH370 Report is a Smokescreen," April 18, 2014, http://jeffwise.net/2014/04/18/slate-why-inmarsats-mh370-report-is-a-smokescreen/.

497 *The People's Trust-Malaysia*, "#MH370—SOS distress call was sent from the plane—China Times," May 7, 2014, http://peoplestrustmalaysia.wordpress.com/2014/05/07/

mh370-sos-distress-call-was-sent-from-the-plane-china-times/ — citing *China Times*, March 8, 2014, http://www.chinatimes.com/ realtimenews/20140308003502-260401.

498 Gunson, "Evidence of a Massive Hoax by the Malaysian Government".

499 Gunson, private message to author, September 8, 2014.

500 Belzer & Wayne, *Dead Wrong: Straight Facts on the Country's Most Controversial Cover-Ups.*

501 John Goglia, "Four Hour Gap In Search And Rescue Revealed By MH370 Preliminary Accident Report," May 1, 2014, *Forbes*, www.forbes.com/sites/johngoglia/2014/05/01/four-hour-gap-in-search-and-rescue-revealed-by-mh370-preliminary-accident-report/.

502 James Corbett, "How to Steal an Airplane: From 9/11 to MH370," March 18, 2014, http://www.corbettreport.com/how-to-steal-an-airplane-from-911-to-mh370/; *Osnet Daily*, "BOMBSHELL: BOEING warned of Cyberjacking, March 11, 2014, http://osnetdaily.com/2014/03/bombshell-boeing-warned-of-cyberjacking-china-deploys-satellites-to-search-for-flight-370/; Deborah Dupre, "Malaysia jet hidden by Electronic Weaponry? 20 EW defense-linked passengers," March 9, 2014, *Examiner.com*, http://www.examiner.com/article/malaysia-jet-hidden-by-electronic-weaponry-20-ew-defense-linked-passengers; *Osnet Daily*, "REVEALED: Flight 370 Communication Transcript strengthens Cyberjacking theory," March 22, 2014, http://osnetdaily.com/2014/03/revealed-flight-370-communication-transcript-strengthens-cyberjacking-theory/.

503 Ralph Waldo Emerson, *The Conduct of Life* (Ticknor and Fields: 1860).

504 *Malaysia 370: The Plane That Vanished*, Documentary produced by Smithsonian Channel.

505 Ibid.

506 Ibid.

507 Ibid.

508 Ibid.

509 Garrison, "Media Cover Up? Jumbo Jet Witnessed Heading Toward Diego Garcia on March 8".

510 Chang, "MH370- A Sinister Tragedy in the Fog of Coincidence? Some strange parallels with catastrophic consequences".

511 Helton, "What's Behind the Disappearance of Malaysian Airliner MH370? Part One:".

512 Michael Ross, "Michael Ross: Was Malaysia Airlines Flight MH370 headed for Asia's Twin Towers?," March 10, 2014, *National Post*, http://fullcomment.nationalpost.com/2014/03/10/michael-ross-was-malaysia-airlines-flight-mh370-headed-for-asias-twin-towers/.

513 Bruce Baker, "Malaysia Airlines Flight 370 'did not crash': Skeptic CEO offers shocking theory," October 13, 2014, *examiner.com*, http://www.examiner.com/article/malaysia-airlines-flight-370-did-not-crash-skeptic-ceo-offers-shocking-theory.

514 Adams, "Six important facts you're not being told about lost Malaysia Airlines Flight 370," emphasis in original.

515 Dr. Kevin Barrett, "MH370: 9/11-style false flag gone awry?," March 31st, 2014, *Veterans Today*, www.veteranstoday.com/2014/03/31/mh370-2/.

516 Stone, "Don't believe the black box ping theory," emphasis in original.

517 Darren Boyle, "'MH370 was under control, probably until the very end,': Emirates CEO Sir Tim Clark reveals doubts over official view of missing airliner's fate," October 10, 2014, *Daily Mail*, www.dailymail.co.uk/news/article-2788599/mh370-control-probably-end-emirates-ceo-sir-tim-clark-reveals-doubts-official-view-missing-airliner-s-fate.html.

518 Ibid.

519 Ibid.

520 Ibid.

521 Ibid.

522 Ibid.

523 Candace Sutton, "'Planes don't just disappear': Former Malaysian Prime Minister accuses CIA of covering up what really happened to flight MH370," May 19, 2014, *DailyMail*, http://www.dailymail.co.uk/news/article-2632447/CIA-knows-missing-Flight-MH370-says-former-Malaysian-PM-Dr-Mahathir.html#ixzz3GSIzVnEm.

524 William Shakespeare, *The Merchant of Venice* (Thomas Heyes: 1600).

525 *About.com*, "Women's History," retrieved August 6, 2014, http://womenshistory.about.com/od/quotes/a/marie_curie.htm.

526 Barry Popik, "Entry from May 22, 2012: 'A foolish faith in authority is the enemy of the truth'," May 22, 2012, *TheBigApple*, http://www.barrypopik.com/index.php/new_york_city/entry/a_foolish_faith_in_authority_is_the_enemy_of_the_truth/.

527 Hutchinson, "Ex-Malaysian prime minister says CIA, Boeing 'hiding' missing airplane".